IAM

INFORMATIONAL ACTUALIZATION MODEL

THE SYNTHESIS: HOW MODERN PHYSICS, CLASSICAL PHILOSOPHY, AND CHRISTIAN THEOLOGY CONVERGE ON A COHERENT VISION OF REALITY

H. W. MAHAFFEY

PRAISE FOR IAM

"In an era increasingly fragmented by hyper-specialization, H. W. Mahaffey's (IAM) The Informational Actualization Model stands as a courageous and necessary work of interdisciplinary synthesis. By grounding the rigorous mathematical landscape of modern cosmology in the deep etymological structures of the ancient world—specifically the creative 'Word' of the Hebrew Davar and the universal logic of the Logos—Mahaffey reconstructs a bridge between science and the human spirit that has been dismantled by centuries of academic 'siloing.'
As a linguist, I find his use of linguistic architecture to navigate the 'fizz of potential' in the early universe to be both conceptually brilliant and structurally sound. This book is a vital invitation for the next generation to reclaim their intellectual inheritance and to begin wondering freely once again."
—Dr. Valentina Sentsova, Professor of Language Mediation and Intercultural Communication, University of Milan; PhD in Pedagogy, Saint Petersburg State University

The Informational Actualization Model

The Synthesis: How Modern Physics, Classical Philosophy, and Christian Theology Converge on a Coherent Vision of Reality

Copyright © 2026 by H. W. Mahaffey

All rights reserved. No part of this book may be reproduced, stored in a retrieval system, or transmitted in any form or by any means—electronic, mechanical, photocopying, recording, or otherwise—without the prior written permission of the author, except for the use of brief quotations in a book review or scholarly journal.

Unless otherwise indicated, Scripture quotations are from The Holy Bible, English Standard Version® (ESV®), copyright © 2001 by Crossway, a publishing ministry of Good News Publishers. Used by permission. All rights reserved. Scripture quotations marked (NASB) are taken from the (NASB®) New American Standard Bible®, Copyright © 1960, 1971, 1977, 1995, 2020 by The Lockman Foundation. Used by permission. All rights reserved. www.lockman.org Scripture quotations marked (NABRE) are taken from the New American Bible, Revised Edition © 2010, 1991, 1986, 1970 Confraternity of Christian Doctrine, Washington, D.C. and are used by permission of the copyright owner. All Rights Reserved. No part of the New American Bible may be reproduced in any form without permission in writing from the copyright owner. Scripture quotations marked (KJV) are taken from the King James Version. Public domain.

Disclaimer:

The ideas and scientific interpretations presented in this book are the author's own and represent a synthesis of current physics, classical metaphysics, and theology. While every effort has been made to ensure the accuracy of scientific and historical information, science is an evolving field. This book is intended for educational and philosophical purposes.

Cataloging-in-Publication Data:

The Informational Actualization Model: The Synthesis/ H. W. Mahaffey. — 1st ed.

(Religion and Science, Cosmology, Metaphysics, Apologetics).

ISBNs:

Hardcover: 979-8-9947108-0-7

Paperback: 979-8-9947108-1-4

eBook: 979-8-9947108-2-1

LCCN: 2026902467

Credits:

Cover Design and Interior Layout by H. W. Mahaffey.

First Edition: 2026 Printed in the United States of America

For permissions or inquiries, contact: hmahaffeyges@gmail.com

*For those who seek truth without fear,
who ask the questions others avoid,
and who refuse to believe that faith and reason
must ever be enemies.*

*And for my children,
so they may know that the universe is not cold,
not silent, and not accidental.
It is the work of the God who says,
"I AM."*

* * *

By faith we understand that the universe was
ordered by the **Word** of God, so that what is visible came
into being through the invisible.
—*Hebrews 11:3 (NABRE)* *

---

* *The capitalization of 'Word' highlights reference to the divine Logos/Christ: See John 1:1-3.*

ACKNOWLEDGMENTS

This work represents a journey that began not in a library, but in wonder—the kind of wonder that refuses to accept that truth must be fragmented, that reason and faith must war, that the cosmos revealed by physics has nothing to say to the soul.

I am profoundly grateful to Dr. Valentina Sentsova, whose expertise in pedagogy and linguistic architecture gave her the clarity to see what this synthesis attempts, and the generosity to say so publicly. Her endorsement means more than she may know, not because it validates the work, but because it confirms that the bridge I've tried to build between ancient wisdom and modern science can bear weight. To have a scholar of her caliber recognize the structural integrity of this approach is a gift I don't take lightly.

I owe a deep debt to the thinkers and traditions that made this work possible: the Hebrew prophets who understood that divine speech creates reality; the Greek philosophers who recognized the Logos pervading all things; the early Church Fathers who synthesized faith and reason without embarrassment; Thomas Aquinas, who showed that Aristotle and Scripture could illuminate one another; the Salamanca School, whose moral philosophy shaped the concept of universal human dignity; Isaac Newton, who devoted more intellectual energy to theology than physics; and the American Founders, who built a republic on the conviction that rights come from God, not government.

I am grateful for my colleagues in the power industry—engineers, operators, and reliability coordinators—who taught me systems thinking in environments where failure has consequences.

The discipline required to coordinate complex, interconnected systems without cascading collapse is not so different from the intellectual care required to synthesize multiple disciplines without category errors. That professional training shaped how I approached this work.

I am thankful for the Christian tradition that preserved not only Scripture but the intellectual rigor to ask hard questions about what it means. I am grateful for the American heritage that insists human dignity is not negotiable, that truth exists to be discovered, and that wonder is not weakness but the beginning of wisdom.

Most of all, I thank my wife and children, whose patience, love, and support made it possible to pursue years of study while managing the demands of work and family life. They never asked me to choose between intellectual depth and daily presence, they trusted that both mattered. This book exists because they believed the journey was worth taking.

All errors, oversights, and failures of clarity are mine alone. Any truth discovered here belongs to the One who is the ground of all being, the eternal I AM, the Logos through whom all things were made, and the Spirit who sustains creation from moment to moment.

ABOUT THE AUTHOR

H. W. Mahaffey is a linguistic architect, independent researcher, and senior reliability coordinator managing hydroelectric operations on the Columbia River. A self-taught polyglot fluent in eight languages, his intellectual journey began with an exhaustive study of ancient Hebrew and Greek—specifically Davar (the creative word-deed), Ruach (spirit/breath), and Logos (universal reason).

These linguistic inquiries led him into a surprising intersection with modern information theory and theoretical cosmology. Over a 30-year career spanning Iceland, the Faroe Islands, Denmark, and the United States—working as a naval air traffic controller, Arctic commercial fisherman, power trader, and systems integrator across all three U.S. power interconnections—Mahaffey pursued disciplined self-study in higher mathematics and physics, becoming intimately familiar with the technical landscape of modern cosmology while managing some of the world's most demanding operational environments.

Recognizing that modern academia often isolates science, philosophy, and theology, Mahaffey looked to historical figures like Thomas Aquinas and Isaac Newton who demonstrated that faith and reason need not oppose each other. He realized the foundations of Western law, justice, and human dignity emerged from such synthesis—an inheritance now being lost to fear of intellectual overlap.

As a husband and father of six, Mahaffey's mission is to reclaim this lost inheritance for the next generation through intellectual scaffolding that allows readers to follow his path without repeating years of raw research. He lives in the Pacific Northwest with his family, where he remains committed to the idea that the highest form of human intelligence is the ability to wonder freely.

PREFACE

This book begins with a simple observation: many of us carry questions that do not fit neatly into any one category.

They may arise from science, from philosophy, from personal experience, or from moments of quiet reflection. Questions about meaning, truth, freedom, consciousness, and why anything exists at all tend to surface whether we ask for them or not. They linger—sometimes gently, sometimes insistently—asking to be taken seriously.

You do not need a particular background to be here. Some readers will come with faith, others with doubt, others with a mixture of both. Some may be returning to questions they once set aside; others may be encountering them for the first time. Wherever you find yourself, this book assumes only curiosity and a willingness to think carefully.

For much of history, wondering about the nature of reality was not seen as a threat to belief or to reason. The same impulse that led people to study the stars, the structure of matter, and the laws of nature also led them to ask deeper questions about meaning and purpose. Wonder was not something to outgrow; it was something to refine.

In recent generations, these ways of thinking have often been separated. Science is frequently asked to explain everything, while questions it cannot answer are quietly set aside. Faith, when it appears at all, is sometimes treated as private or symbolic rather than as a serious way of engaging reality. Many people live with this division without ever being told that another approach is possible.

This book explores that possibility.

It does not ask you to accept claims without reflection or to set aside what you already know. Instead, it invites you to follow the questions where they naturally lead—to see what modern science reveals about the universe, to notice where its explanations reach their limits, and to consider whether older ways of thinking might still speak meaningfully to the world we now understand more clearly.

The chapters move slowly and deliberately. They are meant to be read with patience, not rushed for conclusions. You may find ideas here that resonate immediately, and others that require time. That is part of the process. Honest inquiry rarely unfolds all at once.

If you are reading because you sense that reality is deeper than it appears, that reason and meaning may belong together, or that wonder itself deserves careful attention, then this book is written for you.

You are not expected to arrive with certainty.

Only with openness.

RECLAIMING WONDER

Many people grow up believing they must choose between science and faith, reason and belief, evidence and Scripture. The story is familiar: as our understanding of the world advances, older ways of seeing it quietly fade away. But what if that story is incomplete, and the choice itself unnecessary?

Modern physics reveals a universe that is not chaotic or accidental, but ordered, relational, and deeply intelligible. It is shaped by structure, law, and meaning. This is strikingly consistent with the vision of reality offered by Christian theology: a universe grounded in the eternal I AM, Being itself, the Logos through whom all things are made, and the relational life of the Triune God. Seen this way, science and theology are not rivals competing for authority, but complementary ways of attending to the same reality. When allowed to speak together, they illuminate rather than diminish one another.

What is at stake here is not merely theoretical. We live in a time when assumptions about truth, dignity, freedom, and meaning are quietly shifting. When human dignity is no longer understood as inherent, it becomes vulnerable. When truth is treated as flexible or provisional, it loses its power to ground justice, responsibility, and hope. These changes often unfold subtly, but their consequences shape how societies understand what, and who, matters.

Many of the moral and cultural commitments we now take for granted—freedom of conscience, equality before the law, and the inherent worth of every person—did not arise by chance. They emerged from a long and careful conversation between Greek philosophy, Roman law, Jewish monotheism, and Christian theology. Over centuries, this synthesis helped give rise to universities, hospitals, human rights, the scientific method, and the rule of law. At its center was a transformative conviction: every human being, regardless of status, strength, or ability, bears the image of God and therefore possesses an unshakable dignity.

Whether or not you have thought about it in these terms, you are an heir to this tradition. This book is an invitation to explore that inheritance—to see how insights about the universe, human nature, and meaning itself ultimately converge rather than conflict.

The chapters unfold gradually, each building on the last. The journey begins with science, particularly physics and information theory, moves through philosophy, and then opens fully into theology. Together, these perspectives offer a way of understanding reality that respects the integrity of each discipline while allowing them to inform one another.

The (IAM) Informational Actualization Model's framework does not depend on speculative claims or novel discoveries. Instead, it draws together ideas refined over centuries and invites the reader to see how they form a coherent and meaningful whole. By allowing science, philosophy, and theology to address the questions they are best suited to answer, a broader and more unified vision of reality comes into view.

Written for thoughtful readers, this book aims to be both serious and accessible. No specialized background is required. Whether read privately, shared in conversation, or used as a resource for teaching and reflection—it is designed to invite understanding, dialogue, and wonder.

The structure unfolds in stages, each building on what came before:

Part I: The Universe as Physics Sees It
(Chapter 1–4)

Modern physics reveals a universe with a beginning, governed by rational laws, describable using information theory, fine-tuned for complexity, and directed by an arrow of time. These are empirical facts, not theological assertions. You'll see what cosmology, quantum mechanics, and thermodynamics tell us about the structure of reality.

Part II: The Philosophy Beneath the Physics
(Chapter 5–9)

Physics reaches natural boundaries. It describes how things work but cannot explain why anything exists, why the universe is intelligible, or what consciousness is. These are metaphysical questions requiring metaphysical answers. You'll explore where science stops and philosophy begins, and why both are necessary for a complete picture. You'll explore the other "Why," the one Isaac Newton devoted most of his time to.

Part III: The Informational Actualization Model
(Chapter 10–12)

Here the synthesis comes into focus. Physics, philosophy, and theology converge when each operates within its proper domain. The I.A.M. is not a scientific theory, but rather a framework showing how these disciplines illuminate a single, coherent vision of reality.

Part IV: The Logos—Where Physics Meets Scripture
(Chapter 13–15)

The divine Reason (Logos) through whom all things were made connects modern physics to ancient theology. You'll see how Genesis describes creation as information unfolding, how John 1 presents Christ as the rational structure of reality, and why this matters for understanding both science and faith.

Part V: The Relational God
(Chapter 16–18)

Christian theology reveals God as Trinity, fundamentally relational. This has profound implications for understanding personhood, consciousness, and the nature of reality itself. You'll explore what it means that Being Itself is personal and relational, not abstract and static.

Part VI: Why Christianity Matters for Freedom (Chapter 19–21)

Christianity uniquely grounds the values that define Western civilization: human dignity, individual liberty, and equal justice. You'll trace the intellectual lineage from Aquinas through the Salamanca School to Locke and the American Founding, seeing why these ideas emerged from Christian soil and why secular substitutes have failed to support the same weight.

Part VII: The Human Story

(Chapter 22-23)

What does all this mean for you? This section explores the sermon on the mount as it addresses identity, purpose, love, and the moral structure of reality thus showing how the synthesis illuminates what it means to be human.

Part VIII Questions and Answers

(Questions 1-100)

One hundred specific questions—answered clearly and directly, all addressed with rigor and honesty—integrate modern physics and metaphysics with biblical theology and offer practical applications. Scripture references appear naturally within the text to guide further study.

Part IX: Conclusion

(Chapter 24)

The universe as a personal invitation. Where do we go from here?

Part X: Technical Reference Manual

This is your scholarly safety net. It addresses domain boundaries, methodological clarifications, category errors, and domain separation principles. It explicitly denies the most common positions often misattributed to interdisciplinary writings, makes clear the most common misunderstandings with clarifications. It includes anticipated objections and scholarly responses, multiple cross-discipline translation tables with integration guidelines for interdisciplinary interpretation. It also establishes explicit methodological rules for combining insights across different fields while preserving their boundaries. It ends with a comparative frameworks section that compares this model to other modern theories that may appear to similar and clarifies how they differ.

How to Use This book:

This book can be read straight through, used as a study guide, or broken into parts for group discussion. Technical terms and complex concepts are explained as they arise, and you'll find a Glossary at the back for quick reference.

If you're reading alone:

Work through it chapter by chapter without skipping any the first time reading this book. If a more detailed explanation is needed, consult the Technical Reference Manual (Part X).

If you're studying in a group:

Each part works as a unit for discussion. The questions in Part VIII can guide conversation. The Technical Reference Manual (Part X) settles disputes and clarifies confusion.

If you're skeptical:

Start with Part I to see what physics actually reveals. Then move to Part II to understand why physics alone can't answer every question. Reserve judgment on the theological claims until you've seen the full argument.

If you're a believer looking for intellectual grounding:

This book shows that your faith is not anti-intellectual. It demonstrates how Scripture, philosophy, and science form a coherent whole when properly understood.

How to Use the Chapter Summaries:

Each chapter concludes with a summary box containing:

Key Concepts: The main ideas just presented

What This Means: The practical significance or takeaway.

Important Terms & Concepts: With exact references to Part X for detailed definitions.

Looking Ahead: How this chapter connects to future chapters.

Technical Reference: This points to specific Part X sections to help add clarity when needed.

These summaries serve multiple purposes:

Before reading: Preview what's coming to orient yourself

After reading: Consolidate what you've learned

For review: Quick reference to recall the argument's structure

For discussion: Focus group conversations on key concepts

For navigation: Find where concepts are defined in Part X

When you encounter unfamiliar terms or need deeper clarification, follow the Part X references in the summaries. Learning to use Part X as your technical companion will greatly enhance your understanding of the interdisciplinary synthesis.

What This Book Is—and What It Isn't:

This book does not set out to prove God's existence through physics, nor does it treat the Bible as a science textbook. It does not claim that information itself is conscious, that the universe is a computer simulation, or that theology can be reduced to equations or scientific models.

Instead, this book invites you to notice something quieter and deeper: coherence. It shows that the universe described by modern physics can stand in harmony with the God revealed in Scripture, and that this harmony points toward a deeper unity of truth.

Why Metaphysics? The Path Aquinas Walked:

You might wonder: why does this book move from physics to philosophy before reaching theology? Why not go directly from cosmology to God?

The answer lies in a methodological insight that's over 800 years old, and remains as relevant today as it was in the 13th century.

When Thomas Aquinas encountered Aristotle's philosophy, he faced a challenge similar to the one we face now: how to integrate the best available knowledge of the natural world with Christian theology without collapsing one into the other or forcing them into false conflict.

Aquinas didn't try to make Aristotle's physics 'prove' theology, nor did he reject Aristotle's insights as incompatible with faith. Instead, he recognized that metaphysics—the philosophical study of being, causality, and existence itself—was the necessary bridge.

Physics tells us **how** the natural world behaves. Theology reveals **who** God is and what He has done. But between these two, there's a gap that neither physics nor theology alone can fill: the question of **why** anything exists at all.

That's where metaphysics comes in. It asks questions like:

Why is there something rather than nothing?

What does it mean for something to exist?

What is the relationship between essence and existence?

What kind of being could be the ground of all other beings?

These aren't physics questions (they can't be tested in a lab) and they aren't purely theological questions (they don't require Scripture to formulate). They're philosophical questions—and they're unavoidable if you want to move responsibly from science to theology.

Aquinas used Aristotelian metaphysics as his bridge. The concepts he worked with—act and potency, essence and existence, the four causes, being and becoming—provided the intellectual tools to show how the God of classical theism could be understood as the necessary ground of a contingent, changeable universe.

That synthesis didn't just produce academic philosophy. It gave rise to the intellectual framework that shaped Western universities, moral philosophy, natural law theory, and eventually the concepts of human dignity and inalienable rights that undergird free societies today.

This book follows the same path, not because Aquinas was infallible, but because the structure of reality itself demands it. You cannot responsibly move from 'the Big Bang happened' to 'therefore, the God of Scripture exists' without doing the metaphysical work in between.

So when you reach Part II and encounter terms like 'contingent being,' 'necessary being,' 'essence,' and 'existence,' don't see them as unnecessary complications. See them as the tools that let you **cross the bridge** from **how** the universe works to **why** there is a universe at all.

You're not being asked to become a medieval philosopher. You're being invited to walk an intellectual path that has proven, over centuries, to be the most rigorous and honest way to hold science and faith together without forcing them into false opposition.

The chapters ahead will ask you to think carefully and patiently. They invite you to respect the boundaries between disciplines, to sit with unanswered questions, and to look for connections where conflict is often assumed. As the journey unfolds, you will encounter a vision of reality that is intellectually serious, scientifically grounded, philosophically coherent, and theologically faithful.

Along the way, you may discover why many of history's greatest thinkers never felt forced to choose between faith and reason—and why you don't have to either.

You'll begin to see a universe that is rational because it flows from the Logos, relational because it reflects the Trinity, and intelligible because it bears the imprint of the divine mind. And you'll see that you, as a conscious being who asks "why," are not standing outside this story, but are part of the meaning the universe itself is expressing.

Welcome to the **bridge**—to the **synthesis**—to the
(IAM) Informational Actualization Model.

PART ONE
THE UNIVERSE AS PHYSICS SEES IT

CHAPTER ONE
THE BEGINNING OF ALL THINGS

*I*n the beginning, there was a beginning. This is not poetic language but scientific conclusion. In one of the most profound discoveries of the twentieth century, cosmologists traced the observable universe back to an early hot, dense state roughly 13.8 billion years ago.[1]

The evidence indicates that space, time, matter, and energy all came into being at that specific moment. The universe hasn't always existed. Before that moment, there was no "before"—even time itself began with the universe. To ask what happened before the Big Bang is to ask a question without meaning, like asking what is north of the North Pole.

The Big Bang is not an event that occurred within space and time. It is the origin of space and time itself.[2]

This discovery overturned centuries of scientific consensus. For most of human history, the prevailing view was that the universe is eternal. Ancient Greek philosophers argued for an unchanging cosmos that had always existed and always would. Even into the early twentieth century, most physicists assumed the universe was static, neither expanding nor contracting, simply existing in a stable equilibrium.

Albert Einstein himself, when he formulated general relativity in 1915, found that his equations predicted a dynamic universe, either expanding or collapsing. Rather than accept this result, he introduced a cosmological constant (Λ) to force the universe to remain static. He later reportedly called this his "greatest blunder."[3]

The first hints that the universe might not be static came from observations of distant galaxies. In the 1920s, astronomer Edwin Hubble measured the light from galaxies far beyond our own Milky Way and noticed something extraordinary. The light was redshifted, meaning its wavelength was stretched toward the red end of the spectrum. This redshift indicated that galaxies were moving away from us. And the farther away a galaxy was, the faster it was receding.[4]

This relationship—now known as Hubble's law or the Hubble-Lemaître law[5]—suggested that the universe itself is expanding. If the universe is expanding, then it must have been smaller in the past. Trace the expansion backward in time, and you reach a point where all matter and energy were compressed into an unimaginably dense and hot state. This is the Big Bang. The term was coined somewhat mockingly by physicist Fred Hoyle, who favored an alternative steady-state model. But the name stuck, and the evidence accumulated. The Big Bang is not a theory about an explosion in space. It is a theory about the expansion of space itself, carrying galaxies along with it like raisins in rising bread dough.

Three main lines of evidence support the Big Bang model, and together they make the case overwhelming.

First, the expansion of the universe, confirmed by countless observations of redshifted galaxies.

Second, the cosmic microwave background (CMB)—a faint glow of radiation filling all of space—the remnant heat from the early universe.[6]

Third, the abundance of light elements like hydrogen and helium, which matches the predictions of Big Bang nucleosynthesis,[7] the process by which the first atomic nuclei formed in the minutes after the beginning.

The cosmic microwave background (CMB) is perhaps the most striking piece of evidence. In 1964, physicists Arno Penzias and Robert Wilson detected an unexpected signal while working with a radio antenna. No matter where they pointed the antenna, they picked up a persistent background noise. At first, they thought it was interference or equipment malfunction. They even cleaned pigeon droppings from the antenna, thinking that might be the source. But the signal remained.

It turned out to be the afterglow of the Big Bang—radiation that has been traveling through space for nearly 14 billion years, cooled by the expansion to a temperature of 2.7 Kelvin (about -270°C). Remarkably, this discovery confirmed a prediction made decades earlier by George Gamow and colleagues, who had calculated that if the universe began in a hot, dense state, it should have left precisely this kind of thermal signature.[8]

The CMB is remarkably uniform. In every direction, the temperature is almost exactly the same, varying by only one part in 100,000. This uniformity is evidence that the early universe was extremely smooth and homogeneous. But there are tiny fluctuations small variations in temperature corresponding to regions of slightly higher or lower density. These fluctuations are the seeds from which galaxies and galaxy clusters eventually formed.

The pattern of these fluctuations has been measured with extraordinary precision by three generations of satellites: COBE (1989-1993), WMAP (2001-2010), and Planck (2009-2013).[9] The match between observation and Big Bang predictions is striking: the model's parameters fit the data to better than 1% precision.

The abundance of light elements provides the third pillar of evidence. In the first few minutes after the Big Bang, when the universe was still incredibly hot and dense, nuclear reactions occurred that fused protons and neutrons into the lightest atomic nuclei: hydrogen, helium, and trace amounts of lithium and beryllium. The relative amounts of these elements depend on conditions like temperature, density, and the ratio of protons to neutrons. When physicists calculate what should have formed under the conditions predicted by the Big Bang model, the results match the observed abundances throughout the universe. About 75 percent hydrogen, 25 percent helium, and tiny traces of other light elements. This agreement is not approximate. It is precise. No other model of cosmic origins has come close to explaining this pattern.

These three lines of evidence—cosmic expansion, the microwave background, and primordial nucleosynthesis—together establish that the universe had a beginning. The Big Bang is not speculation. It is the best-supported theory in all of cosmology. Every prediction the model makes has been confirmed by observation. The universe is not eternal. It is not static. It began, and it has been expanding and evolving ever since.

But the Big Bang raises as many questions as it answers. What caused the universe to begin? Why did it start in such a low-entropy, highly ordered state? Why are the laws of nature what they are? Why do the fundamental constants have the values they have? These are questions that physics observes but cannot answer. They point beyond the domain of empirical science toward philosophy and theology.

Consider the initial state of the universe. At the moment of the Big Bang, the universe was not in a state of maximum entropy or disorder. It was extraordinarily ordered. The distribution of matter and energy was smooth and uniform. The gravitational degrees of freedom were barely activated.

This is profoundly puzzling because entropy tends to increase over time. If you start with a low-entropy state and let the system evolve, it will naturally move toward higher entropy. But why did the universe start in such a low-entropy state in the first place?

The physicist Roger Penrose calculated the odds of the initial conditions occurring by chance. The probability is approximately 1 in $10^{\wedge}$ ($10^{\wedge}123$)—a number so large it defies comprehension.[10] To put this in perspective: there are only about $10^{\wedge}80$ atoms in the observable universe. The precision required for the universe's initial conditions dwarfs even that enormous number by factors we lack language to express.

The universe is also fine-tuned. The fundamental constants of nature—the strength of gravity, the electromagnetic force, the mass of the electron, the cosmological constant—all fall within extraordinarily narrow ranges that permit the existence of complexity and life.

Consider just one example: the cosmological constant. If it were larger by even one part in 10^{120}, the universe would have expanded too rapidly for galaxies to form. If it were smaller by a comparable amount in the opposite direction, the universe would have collapsed before stars could ignite.[11] Change any of these constants even slightly, and the universe becomes inhospitable to structure of any kind. Stars do not form. Atoms do not exist. Chemistry does not happen. Life is impossible.

The fine-tuning is not limited to one or two parameters. It appears across dozens of independent constants and initial conditions. The universe seems calibrated for the possibility of physical complexity.

Some physicists propose a multiverse to explain this fine-tuning. If there are countless universes with different constants and laws, then we should not be surprised to find ourselves in one where the conditions permit our existence. We could not observe a universe where life is impossible because we would not be there to observe it. This is called the anthropic principle.

But the multiverse hypothesis, while intriguing, is not itself testable through empirical observation. We cannot observe other universes. We cannot measure their constants or confirm their existence. The multiverse may be a coherent theoretical framework within certain cosmological models,[12] but it functions more as metaphysical interpretation than empirical science. Whether this represents a genuine explanation or simply relocates the question (why does a multiverse-generating mechanism exist?) remains actively debated.

Another mystery concerns the laws of nature themselves. Why does the universe obey mathematical laws? Why is nature governed by principles like conservation of energy, symmetry, and causation?

Physics describes these laws and uses them to make predictions, but it does not explain why the laws exist or why they take the form they do. The laws of nature are not logically necessary. We can imagine universes with different laws. Yet our universe follows this particular set of rules with astonishing precision. Why?

The intelligibility of the universe is another puzzle. Why can human minds, which evolved to survive on the African savanna, understand the behavior of galaxies, black holes, and subatomic particles? The physicist Eugene Wigner called this "the unreasonable effectiveness of mathematics in the natural sciences." Mathematics is a product of human thought, yet it describes the cosmos with uncanny accuracy. This correspondence between the structure of our minds and the structure of reality is not self-explanatory. It suggests that the universe is rational in a deep sense, ordered in a way that makes it knowable.

Why does the universe exist? Why did it begin in a low-entropy state? Why is it fine-tuned? Why does it obey laws? Why is it intelligible? These are not scientific questions in the sense that they can be answered by further experiments or observations. They are metaphysical questions. They ask about the ground of being, the source of order, and the reason for rationality.

Science presupposes existence, order, and rationality. It describes their structure with extraordinary precision. But it does not, and by its methodology, cannot explain why these conditions hold. That requires a different mode of inquiry: philosophy and theology.

The next chapter explores one of the most profound applications of information theory in cosmology: understanding the universe through the lens of meaningful distinctions. What does it mean to say the universe is fundamentally informational? And how does this insight connect to the deeper questions physics leaves unanswered?

CHAPTER REVIEW:
KEY CONCEPTS:
- **The universe had a temporal beginning ~13.8 billion years**

ago - Modern cosmology demonstrates that space, time, matter, and energy all originated together in what we call the Big Bang.

- **This is not an event IN space and time, but the origin OF space and time** - There is no "before" the Big Bang in a physical sense; asking what happened before is like asking what's north of the North Pole.

- **Three lines of evidence converge:** cosmic expansion (Hubble-Lemaître law), cosmic microwave background radiation, and primordial nucleosynthesis all confirm the Big Bang model.

- **The universe began in an extraordinarily low-entropy, highly ordered state** - The probability of this initial configuration occurring by chance is approximately 1 in $10^{10^{123}}$.

WHAT THIS MEANS:

Physics describes HOW the universe began and evolved but cannot answer WHY anything exists at all. This distinction between empirical description and metaphysical explanation sets up the need for philosophical and theological inquiry in later chapters. The temporal beginning points to questions that lie beyond physics itself.

IMPORTANT TERMS & CONCEPTS:

- **Big Bang Model** → Part X, Section 1.3 (Domain Boundaries: Physics)

- **Entropy** (thermodynamic) → Part X, Section 1.4 (Information Theory)

- **Fine-tuning** → Part X, Section 1.3 (Cosmic Fine-Tuning)

- **Low-entropy initial state** → See Chapter 4 for full treatment

- **Cosmological Constant** → Part X, Section 1.3 (Foundations of Cosmology)

LOOKING AHEAD:

- Chapter 2 explores how information theory provides a framework for understanding the universe's structure

- Chapter 4 examines why the low-entropy beginning matters for the arrow of time

- Chapter 5 addresses why physics alone cannot answer questions about existence itself

- Part III (Chapters 10-12) shows how this beginning points toward a necessary ground of being

TECHNICAL REFERENCE:

- For clarification on what "beginning of time" means physically: Part X, Section 1.3 (Singularity - Mathematical)
- For domain separation between physics and metaphysics: Part X, Section 1 (entire section)
- For what this model does NOT claim about the Big Bang: Part X, Section 3

---

1. Planck Collaboration (2020). "Planck 2018 Results VI: Cosmological Parameters." *Astronomy & Astrophysics* 641, A6.
2. In standard Big Bang cosmology, what existed "before" the earliest time described by the model is not a physically well-defined question, as time is part of the system whose history the model traces.
3. The "greatest blunder" attribution comes from George Gamow's autobiography *My World Line* (1970), though the exact quote's authenticity is debated among historians. Regardless of whether Einstein used these precise words, his introduction of the cosmological constant to achieve a static universe—later abandoned when Hubble's observations showed expansion—represents one of physics' most famous conceptual reversals.
4. Hubble, E. (1929). "A Relation between Distance and Radial Velocity among Extra-Galactic Nebulae." *Proceedings of the National Academy of Sciences* 15(3), 168–173.
5. In 2018, the International Astronomical Union voted to rename this the "Hubble-Lemaître law" in recognition that Belgian priest-physicist Georges Lemaître proposed the expansion of the universe in 1927, two years before Hubble's observational confirmation.
6. Penzias, A. A., & Wilson, R. W. (1965). "A Measurement of Excess Antenna Temperature at 4080 Mc/s." *Astrophysical Journal* 142, 419–421.
7. For a comprehensive technical review of Big Bang nucleosynthesis predictions and observational constraints, see Particle Data Group: "Big-Bang Nucleosynthesis" by Fields, Molaro, & Sarkar (2015).
8. Gamow, G. (1948). "The Evolution of the Universe." *Nature* 162, 680–682. Gamow's prediction, initially dismissed by many physicists who favored a steady-state universe, was vindicated by Penzias and Wilson's accidental discovery.
9. The most recent and precise measurements come from the Planck satellite. See Planck Collaboration (2020), "Planck 2018 Results VI: Cosmological Parameters." *Astronomy & Astrophysics* 641, A6.
10. Penrose, R. (2004). *The Road to Reality: A Complete Guide to the Laws of the Universe.* Knopf, pp. 726–732. While the exact value depends on certain assump-

tions about quantum gravity (which remains incomplete), even conservative estimates yield astronomically small probabilities.

11. For a comprehensive survey of fine-tuning across multiple physical parameters, see Lewis, G. F., & Barnes, L. A. (2016). *A Fortunate Universe: Life in a Finely Tuned Cosmos*. Cambridge University Press.
12. For multiverse proposals within inflationary cosmology, see Guth, A. H. (1997). *The Inflationary Universe*. Basic Books. For philosophical critiques, see Ellis, G. F. R. (2011). "Does the Multiverse Really Exist?" *Scientific American* 305(2), 38–43.

CHAPTER TWO
THE UNIVERSE AS INFORMATION

*I*n the middle of the twentieth century, a Bell Labs engineer named Claude Shannon developed a mathematical theory that would transform how we understand communication, computation, and eventually reality itself. Shannon was not thinking about cosmology or quantum mechanics when he worked out his theory of information. He was trying to solve a practical problem: how to transmit messages efficiently over noisy telephone lines.[1] But the framework he created turned out to describe something far more fundamental than telegrams and phone calls. It described the structure of distinction itself.[1]

Shannon's key insight was that information is not about meaning. It is about reduction of uncertainty. When you receive a message, you learn something you did not know before. The amount of information in that message is proportional to how much your uncertainty decreases. If someone tells you the outcome of a coin flip, you gain one bit of information because you had two possible outcomes and now you know which one occurred. If someone tells you the result of rolling a six-sided die, you gain more information because there were more possibilities to begin with. Information, in Shannon's framework, is measured in bits—

binary distinctions between yes and no, zero and one, true and false.

This might seem abstract, but it has profound implications. It means that at the most basic level, reality can be described in terms of distinctions. To specify the state of any system, you need to answer a series of yes-or-no questions. Is the particle here or there? Is the spin up or down? Is the atom in this energy level or that one? Every physical system can be characterized by the distinctions that define it. And the more distinctions required to fully specify a system, the more information it contains.

This framework has proven remarkably powerful. It underlies all of modern computing. Every calculation performed by a computer, every image displayed on a screen, every word processed by software ultimately reduces to sequences of bits—strings of zeros and ones encoding distinctions. But the reach of information theory extends far beyond technology. It has become a fundamental tool in physics, appearing in contexts from quantum mechanics to cosmology to black hole thermodynamics.

One of the most striking applications of information theory in physics emerged from the study of black holes. In the 1970s, physicist Jacob Bekenstein proposed that black holes have entropy—a measure of disorder or hidden information. This was a radical idea because black holes were thought to be the simplest objects in the universe, completely characterized by just three numbers: mass, charge, and angular momentum. How could something so simple have entropy?[2]

Bekenstein's insight was that while black holes appear simple from the outside, they can form from the collapse of matter in countless different configurations. A black hole could form from a collapsing star, from compressed gas, from pure radiation, or from any combination of ingredients, and as long as the mass, charge, and spin were the same, the resulting black hole would be indistinguishable from any other. All the information about what went into making the black hole is hidden behind the event horizon. That hidden information is what Bekenstein identified as entropy.

Stephen Hawking, building on Bekenstein's work, calculated the precise value of this entropy. The Bekenstein-Hawking formula states that the entropy of a black hole is proportional to the area of its event horizon, not its volume. In natural units, the entropy equals one-fourth of the horizon area measured in Planck units. This discovery was profound. It suggested that the maximum amount of information that can be stored in any region of space is determined not by the volume of that region but by the area of its boundary. Information, in a deep sense, seems to be a surface phenomenon rather than a volumetric one.[3]

This insight gave rise to what is now called the holographic principle. The principle states that all the information contained within a volume of space can be encoded on the boundary of that volume, much like a hologram encodes a three-dimensional image on a two-dimensional surface.[4]

This is not merely a metaphor. In certain well-understood contexts—particularly in the AdS/CFT correspondence developed in string theory—the holographic principle is a precise mathematical statement about the relationship between gravity in one space and quantum field theory on the boundary of that space. The principle suggests that what we perceive as three-dimensional reality might be describable in terms of information encoded on a two-dimensional surface.[5]

The holographic principle remains an active area of research, and its full implications are not yet understood. But it has already reshaped how physicists think about space, information, and the fundamental structure of reality. It suggests that space itself might not be fundamental but emergent, arising from more basic informational or quantum mechanical structures. This does not mean the universe is a simulation or that reality is "just information" in some literal sense. It means that information-theoretic language provides a useful and perhaps necessary framework for understanding the deepest layers of physical law.

Another cornerstone of information theory in physics is Landauer's principle, formulated by Rolf Landauer in 1961.[6]

Landauer showed that erasing information has a thermodynamic cost. To erase a single bit of information—to reset a memory element from one or zero back to a neutral state—requires a minimum amount of energy, proportional to the temperature of the system. This principle establishes a fundamental link between information and thermodynamics. Information is not an abstract concept floating above physical reality. It is embedded in the states of physical systems, and manipulating it requires real physical work.

Landauer's principle has practical implications for computing. Every time a computer erases data, it must dissipate energy as heat. This is why computers get warm when they run. The more information processed and the more bits erased, the more heat generated. As computers become faster and more powerful, the thermodynamic cost of computation becomes increasingly significant. Engineers designing quantum computers and other advanced systems must account for this fundamental limit. You cannot process information without affecting the physical world.

But Landauer's principle also has deeper theoretical significance. It shows that information is physical. The distinction between information and matter, between abstract data and concrete substance, is not absolute. Information is encoded in the configuration of physical systems, and changing those configurations requires energy. This suggests that at the most fundamental level, distinctions—whether between particle positions, quantum states, or any other measurable properties—are not merely conceptual. They are features of reality itself.

Quantum mechanics adds another layer to this picture. In the quantum realm, systems do not always have definite properties. A quantum particle can exist in a superposition of multiple states simultaneously. It is not that we do not know which state the particle is in. The particle genuinely does not have a definite state until it interacts with something else.[7] Only when a measurement or interaction occurs does the superposition collapse into one definite outcome. Before that moment, the particle exists in a state of potentiality, with multiple possibilities coexisting.

This quantum indeterminacy has profound implications for information. Before a measurement, the information about the system is not merely hidden. It does not yet exist in a definite form. The act of measurement brings information into being by selecting one outcome from many possibilities. This process, called decoherence, is irreversible.[8] Once the environment has recorded the outcome, there is no going back to the original superposition. The information has been generated, and it persists in the correlations between the system and its surroundings.

This suggests that information in quantum mechanics is not just a passive description of what already exists. It is tied to the process by which definite states emerge from quantum potentials. Every interaction, every decoherence event, every measurement generates information that becomes part of the physical record. The universe is constantly creating information as quantum systems interact and as possibilities become actualities. This is not a claim about consciousness or observers causing reality. It is a statement about the structure of quantum mechanics itself.

The connection between quantum mechanics and information has led to the development of quantum information theory, a field that explores how quantum systems can be used to encode, process, and transmit information in ways impossible for classical systems. Quantum entanglement, for instance, allows particles to be correlated in ways that defy classical explanation.[9] Two entangled particles can be separated by vast distances, yet a measurement on one instantaneously affects the probabilities for measurements on the other. This is not communication faster than light—no information is transmitted—but it is a form of correlation that cannot be described by any local classical mechanism. Entanglement is now understood as a resource, a kind of quantum information that can be harnessed for tasks like quantum computing, quantum cryptography, and quantum teleportation.

All of this paints a picture of a universe that is deeply structured by distinctions, correlations, and patterns. Physical systems encode information in their states. Interactions generate new information. The laws of physics constrain how information can be transformed and transmitted. And the very notion of a physical state—whether of a particle, a field, or the universe as a whole—can be described in information-theoretic terms.

But it is crucial to understand what this does and does not mean. When physicists say the universe can be described using information theory, they are not claiming that information is more fundamental than matter or energy. They are saying that the mathematical language of information provides a powerful and sometimes necessary tool for describing physical processes. Information theory is a framework, a way of thinking about distinctions, correlations, and constraints. It does not replace the equations of physics. It complements them.

Nor does describing the universe in information-theoretic terms mean that the universe is a computer or a simulation. Computers process information according to algorithms. The universe evolves according to physical laws. These are different processes. A quantum field does not "compute" its evolution. It simply evolves according to the Schrödinger equation. The fact that we can describe that evolution in terms of information does not make the field any less real or physical. It simply means that the language of information is useful for understanding the structure of quantum mechanics.

Similarly, the holographic principle does not mean we live in a hologram. It means that in certain gravitational contexts, the information content of a region can be fully captured by data on the boundary. This is a statement about the mathematical structure of spacetime and quantum field theory, not a claim about the illusory nature of reality. Space, particles, forces—all of these are real. The holographic principle is a tool for understanding the relationships between them, not a denial of their existence.

What information theory does reveal is that the universe is structured by meaningful distinctions. To specify the state of anything, you must distinguish it from other possibilities. To describe any process, you must track which distinctions change and which remain constant. This is true whether you are describing a coin flip, a quantum measurement, or the evolution of the cosmos. The universe is not a featureless continuum. It is articulated, differentiated, ordered. And that order can be characterized in terms of the distinctions that define it.

This observation resonates with insights from philosophy and theology. Classical metaphysics has always recognized that being is not undifferentiated existence. To be is to be something rather than something else. A thing has an essence—a whatness that distinguishes it from other things. It has properties, relationships, and boundaries. Reality is structured by form, and form is what makes distinctions possible. Information theory, in a sense, is the mathematical language for describing that structure.

Theology goes further. If the universe is created by the Word—if God speaks and creation comes into being—then creation unfolds through the establishment of distinctions. Genesis portrays this explicitly: light is separated from darkness, waters above from waters below, sea from land, day from night. Each creative act introduces a new distinction, and each distinction generates order. The formless void becomes a cosmos because God articulates differences, names things, and assigns them their places. The language of divine speech is the language of meaningful distinction. And that is precisely what information theory describes in mathematical terms.

This does not mean Genesis predicted information theory or that physics proves Scripture. It means that both describe reality using the concept of meaningful distinction. One does so theologically, the other mathematically. But they are not contradictory. They are complementary perspectives on the same fundamental truth: the universe is ordered, and that order arises from the establishment of differences.

The universe revealed by modern physics is not a brute collection of particles obeying blind forces. It is a structured, relational, intelligible reality characterized by patterns, correlations, and distinctions. Information theory provides a language for describing that structure. It shows that the universe is, in a deep sense, knowable because it is organized. And that knowability points toward something beyond physics itself: the question of why there is order at all, why distinctions are stable, and why the universe is structured in a way that minds can grasp. Those questions belong to philosophy and theology. But physics sets the stage for asking them.

The next chapter explores one of the most profound applications of information theory in cosmology: the holographic principle and its implications for the nature of space, time, and reality itself.

CHAPTER REVIEW:
KEY CONCEPTS:

- **Shannon information theory measures reduction of uncertainty through distinctions** - Information is fundamentally about meaningful differences, not about meaning itself

- **Black holes have entropy proportional to their surface area, not volume** (Bekenstein-Hawking formula: $S = A/4$ in Planck units) - This revolutionary insight connects information, gravity, and thermodynamics

- **Landauer's Principle establishes that erasing information requires energy** - Information is not abstract but physically embedded in reality

- **The universe can be described using information-theoretic language** - Distinctions, correlations, and boundaries characterize physical systems at the most fundamental level

WHAT THIS MEANS:

The universe is structured by meaningful distinctions at every level. Information theory provides a mathematical framework for describing this structure, but information itself is NOT conscious, purposeful, or semantic. The informational nature of reality is compatible with theological claims about creation through divine speech (establishing distinctions), but this is structural analogy, not proof.

IMPORTANT TERMS & CONCEPTS:

- **Shannon Information** → Part X, Section 1.4 (Shannon Information/Shannon Entropy)

- **Black Hole Entropy** → Part X, Section 1.4 (Black Hole Information Problem)

- **Bit** (as unit of information) → Part X, Section 1.4 (Bit - Binary Digit)

- **Landauer's Principle** → Part X, Section 1.4 (Landauer's Principle)

- **Information vs. Meaning** → Part X, Section 1.4 (Information - General Definition)

LOOKING AHEAD:

- Chapter 3 develops the holographic principle showing information encodes on boundaries

- Chapter 13 connects Genesis creation narrative to information as meaningful distinction

- Chapter 14 explores the Logos (divine Word/Reason) as the source of informational order

- The synthesis in Part III shows how informational structure points beyond physics

TECHNICAL REFERENCE:

- For the crucial distinction between information and meaning: Part X, Section 1.4 (Information - General Definition)

- For what "informational universe" does NOT mean: Part X, Sections 3 & 4 (Common Misunderstandings)

- For analogy vs. identity when connecting information to theology: Part X, Section 7.5 (Analogical Interpretation)

---

1. Shannon, C. E. (1948). "A Mathematical Theory of Communication." *Bell System Technical Journal* 27(3), 379–423. This foundational paper established information theory as a formal discipline, defining entropy in communication systems and proving fundamental theorems about channel capacity and compression
2. Bekenstein, J. D. (1973). "Black Holes and Entropy." *Physical Review D* 7(8), 2333–2346. Bekenstein proposed that black holes must possess entropy proportional to their horizon area to preserve the second law of thermodynamics.

3. Hawking, S. W. (1975). "Particle Creation by Black Holes." *Communications in Mathematical Physics* 43(3), 199–220. Hawking demonstrated that black holes emit thermal radiation and calculated the precise relationship between entropy and horizon area: $S = A/4$ in Planck units.
4. t Hooft, G. (1993). "Dimensional Reduction in Quantum Gravity." arXiv:gr-qc/9310026. Susskind, L. (1995). "The World as a Hologram." *Journal of Mathematical Physics* 36(11), 6377–6396. These papers formulated the holographic principle, proposing that the information content of a spatial region is encoded on its boundary.
5. Maldacena, J. (1998). "The Large N Limit of Superconformal Field Theories and Supergravity." *Advances in Theoretical and Mathematical Physics* 2(2), 231–252. The AdS/CFT correspondence provides the most concrete realization of holography, establishing an exact equivalence between gravitational theories in Anti-de Sitter space and conformal field theories on the boundary.
6. Landauer, R. (1961). "Irreversibility and Heat Generation in the Computing Process." *IBM Journal of Research and Development* 5(3), 183–191. Landauer showed that erasing one bit of information requires a minimum energy expenditure of $kT \ln 2$, where k is Boltzmann's constant and T is temperature.
7. In quantum mechanics, "measurement" refers to any physical interaction that produces a definite outcome, not necessarily observation by a conscious mind. This distinction is crucial to avoid misunderstanding quantum theory as requiring consciousness to determine reality.
8. Zurek, W. H. (2003). "Decoherence, Einselection, and the Quantum Origins of the Classical." *Reviews of Modern Physics* 75(3), 715–775. Decoherence explains how quantum superpositions give way to definite classical states through interaction with the environment.
9. Einstein, A., Podolsky, B., & Rosen, N. (1935). "Can Quantum-Mechanical Description of Physical Reality Be Considered Complete?" *Physical Review* 47(10), 777–780. Bell, J. S. (1964). "On the Einstein Podolsky Rosen Paradox." *Physics* 1(3), 195–200. The EPR paper questioned quantum mechanics' completeness, while Bell's theorem showed that no local hidden variable theory could reproduce quantum predictions, confirming the reality of quantum entanglement.

CHAPTER THREE
THE MATHEMATICS OF A HOLOGRAPHIC UNIVERSE

*B*lack holes should be simple. A star collapses under its own gravity, spacetime warps beyond the point of no return, and what remains is an object completely described by just three numbers: mass, electric charge, and spin. Strip away everything else: the star's chemical composition, its rotation history, the exact sequence of fusion reactions that powered it, and the black hole that forms is identical to any other black hole with the same mass, charge, and spin. This simplicity troubled physicists. If all that information disappears, where does it go?

In the early 1970s, Jacob Bekenstein proposed something remarkable: black holes have entropy.[1] This was not an obvious claim. Entropy measures disorder, the number of hidden configurations a system could have while looking the same from the outside. A box of gas has high entropy because the molecules could be arranged in countless ways while still producing the same temperature and pressure. But black holes were thought to have no internal structure, no hidden configurations. They were defined entirely by those three external parameters. How could they have entropy?

Bekenstein's insight came from thinking about what happens when matter falls into a black hole. A collapsing star carries information—the positions and velocities of its atoms, the quantum states of its particles, the detailed history of its formation. Once that star crosses the event horizon, all that information becomes inaccessible to outside observers. You cannot see inside a black hole. But the information is not destroyed. It must be encoded somehow in the black hole itself. That hidden information is what Bekenstein identified as entropy.

Stephen Hawking refined this idea and calculated the precise value.[2] The result was stunning: the entropy of a black hole is proportional to the area of its event horizon, not its volume. In natural units where fundamental constants are set to one, the formula is elegantly simple: entropy equals one-fourth of the horizon area. This was unexpected. If you think of entropy as measuring stuff hidden inside a volume, you would expect it to grow with the size of that volume. Double the radius of a sphere, and its volume increases eightfold. But black hole entropy does not scale that way. It scales with surface area, not volume. Double the radius, and the surface area—and therefore the entropy—only quadruples.

This pattern hints at something deeper. If the maximum information content of a region is limited by the area of its boundary rather than its interior volume, then perhaps our everyday intuition about space is incomplete. Perhaps the three-dimensional world we experience is not the most fundamental description of reality. Perhaps all the information defining what happens inside a volume can somehow be encoded on the surface surrounding it.

This idea crystallized into what physicists call the holographic principle.[3] The name comes from optical holograms, those two-dimensional surfaces that store three-dimensional images. Shine light through a hologram at the right angle, and the full three-dimensional scene appears. The holographic principle suggests something similar happens in physics: the information content of any region of space can be fully captured by information on its boundary. Three-dimensional reality might emerge from two-dimensional information.

This is not speculation or metaphor. In certain precise mathematical contexts, holography is an exact statement. The clearest example comes from string theory, specifically from a correspondence discovered by Juan Maldacena in 1997.[4] Maldacena showed that a gravitational theory in a particular kind of space—called anti-de Sitter space, or AdS—is mathematically equivalent to a quantum field theory living on the boundary of that space. The boundary theory has no gravity and one fewer spatial dimension. Yet everything happening in the interior—particles moving, black holes forming, spacetime bending—can be completely described by processes on the boundary.

Think about what this means. You have two descriptions that look entirely different. One involves gravity, curved spacetime, and an extra dimension. The other involves quantum fields on a flat boundary with no gravity at all. Yet they describe the same physics. Any question you can ask in the gravitational theory has a precise answer in the boundary theory. The interior is not an illusion, but it is emergent—it arises from the dynamics of something simpler.

This AdS/CFT correspondence, as it is called, has become one of the most powerful tools in theoretical physics. Researchers use it to study exotic states of matter, to understand the behavior of quark-gluon plasmas in particle accelerators, and to explore the nature of quantum entanglement. The correspondence provides a bridge between quantum mechanics and gravity, two frameworks that have resisted unification for nearly a century.

One of the key insights concerns entanglement, that strange quantum phenomenon where measuring one particle instantly affects another, no matter how far apart they are. In the boundary theory, entanglement describes correlations between different regions. In the gravitational interior, those correlations manifest as geometric connections. The more entangled two regions are, the more closely connected they appear in the emergent space. This relationship, captured in the Ryu-Takayanagi formula, suggests something profound: spacetime geometry and quantum entanglement are two sides of the same coin.[5]

Some physicists have taken this further, proposing that space itself is woven from entanglement. In this view, space is not a pre-existing stage on which matter moves. It is an emergent phenomenon arising from quantum correlations. When correlations are strong, space appears smooth and connected. When they weaken, space breaks down. This idea—sometimes summarized as "ER equals EPR," linking Einstein-Rosen wormholes with Einstein-Podolsky-Rosen entanglement—remains speculative, but it points toward a radical rethinking of what space fundamentally is.

For centuries, space was understood as a fixed container. Newton assumed absolute space and time, independent of matter. Einstein showed that space and time are coupled to mass and energy, forming a flexible spacetime that curves and bends. But even in general relativity, spacetime is still fundamental—the arena in which physics unfolds. The holographic principle suggests we may need to go further. Space might not be fundamental at all. It might emerge from something more basic, something informational or quantum mechanical that does not itself have spatial character. Just as temperature emerges from molecular motion and pressure emerges from particle impacts, space might emerge from quantum entanglement.

These ideas remain at the frontier of research. The AdS/CFT correspondence applies to a particular kind of spacetime that does not match our universe. Our cosmos is expanding and appears to have a small positive cosmological constant, making it more like de Sitter space than anti-de Sitter space. Whether holography applies to realistic cosmological settings is an open question. But what is clear is that the principle has fundamentally reshaped how physicists think about information, gravity, and reality itself.

The holographic principle establishes something remarkable: information is not infinitely compressible. There is a maximum amount of information that can be associated with any region, and that maximum is set by the area of its boundary, not its volume. This limit, called the Bekenstein bound, is universal.[6] No process, no matter how clever, can pack more information into a region than this bound allows without causing that region to collapse into a black hole.

This has practical implications. It limits the storage capacity of any physical memory. It constrains the complexity of any physical system. And it ties information directly to gravity. Information has a gravitational footprint. Concentrate too much of it in too small a space, and you warp spacetime to the breaking point.

The connection runs deeper. At the most fundamental level, physical laws might not be about particles moving through space but about information being processed and constrained according to universal principles. This does not mean the universe is a computer—please note that carefully. (The difference between mathematical description and literal mechanism is addressed in detail in Part X, Section 4.) It means that information-theoretic language provides a powerful framework for understanding physics at its deepest layers.

Return now to black holes and the original puzzle about entropy. Black holes represent the maximum concentration of information possible in a given region. They sit at the boundary of the physically possible, where gravity becomes so strong that not even light escapes. The fact that their entropy scales with area rather than volume is not an accident. It reveals something fundamental about how space, information, and gravity interact.

But black holes also pose a deep challenge called the information paradox. When matter falls into a black hole, it seems to vanish. The black hole's exterior is described completely by mass, charge, and spin. All the detailed information about what fell in is hidden behind the event horizon. Hawking showed that black holes slowly evaporate by emitting thermal radiation.[7] Eventually, the black hole disappears entirely. But the Hawking radiation appears to carry no information about what originally fell in. If that information is truly lost, quantum mechanics is violated. One of its bedrock principles is that information is never destroyed.

This paradox has driven decades of research. The resolution is not yet fully understood, but the emerging consensus is that information is not lost. It must be encoded in the Hawking radiation, hidden in correlations too subtle to detect easily but present nonetheless. The mathematics is deep, and the work continues, but the principle stands: quantum mechanics is preserved. Information survives, even in black holes.

The holographic principle, the Bekenstein bound, AdS/CFT, the Ryu-Takayanagi formula—these represent some of the most profound insights in physics over the past half-century. They show that reality is stranger and more subtle than everyday experience suggests. Space might be emergent. Information might be bounded. Gravity might arise from quantum correlations. These insights are still being explored, still being tested in mathematical models, still revealing new implications.

But what they establish is clear. The universe is deeply structured by information in a precise mathematical sense—not metaphorical, but exact. (The distinction between mathematical structure and conscious meaning is crucial and is explained thoroughly in Part X, Section 8.) The distinctions that define physical states, the correlations that link separated systems, the bounds that constrain complexity—all these point to a reality that is fundamentally relational and intelligible.

The universe is not chaos. It is order. That order can be described, understood, and formalized in the language of mathematics. This does not mean we have answered every question. The nature of space, the origin of physical laws, the reason there is something rather than nothing—these remain open. But the progress made over the past century has been extraordinary. We have moved from viewing the universe as solid objects in empty space to understanding it as a web of relationships, correlations, and information governed by elegant mathematical principles.

And that progress raises deeper questions. If space is emergent, what does it emerge from? If information is bounded by area, why area and not volume? If gravity arises from entanglement, what is entanglement at its core? These are not questions physics alone can answer. They lead beyond the laboratory and the equation into the domain of philosophy and theology. They ask not just how the universe works but why it exists at all, why it is ordered, and why it is comprehensible to minds like ours.

The discoveries explored in this chapter do not prove God exists. They do not turn physics into theology. But they do something equally important: they show that the universe revealed by modern physics is compatible with the God revealed in Scripture. The informational structure, the relational character, the mathematical order—these are precisely what we would expect if the cosmos is the work of a rational Creator who speaks reality into being through meaningful distinctions.

Genesis describes creation unfolding through divine speech, through the establishment of differences: light from darkness, sea from land, day from night. Information theory describes reality in terms of distinctions, boundaries, and correlations. The holographic principle reveals that information is not arbitrary but constrained by fundamental limits tied to the geometry of space itself. These insights harmonize. They point to a universe that is not the product of blind chance but of intentional order.

This does not mean ancient authors anticipated modern physics. It means that both Scripture and science describe a reality structured by meaningful distinction. One speaks theologically, the other mathematically. But they converge on a vision of the cosmos as ordered, relational, and intelligible—a cosmos that invites understanding and points beyond itself toward ultimate questions of purpose and origin.

The next chapter turns to another deep aspect of cosmic order: the arrow of time, the nature of entropy, and the process of becoming that gives the universe its direction and history.

CHAPTER REVIEW:
KEY CONCEPTS:
- **The holographic principle states that information in a region encodes on its**
   boundary - The maximum information content scales with surface area, not volume

The Mathematics of a Holographic Universe

- **AdS/CFT correspondence provides an exact mathematical realization** - A gravitational theory in bulk space is equivalent to a quantum field theory on its boundary
- **Quantum entanglement may be the mechanism generating spacetime geometry** (ER=EPR proposal) - Space itself might emerge from informational/quantum correlations
- **The Bekenstein bound sets universal limits on information density** - No region can contain more information than allowed by its boundary area without collapsing into a black hole

WHAT THIS MEANS:

Three-dimensional space may be emergent from more fundamental informational or quantum structures. This does NOT mean the universe is a computer simulation or that matter is an illusion - it means information-theoretic description captures something deep about physical law. The holographic structure shows reality is more subtle than naive materialism suggests.

IMPORTANT TERMS & CONCEPTS:

- **Holographic Principle** → Part X, Section 1.4 (Holographic Information)
- **AdS/CFT Correspondence** → Part X, Section 1.3 (referenced in Holographic Principle discussion)
- **Bekenstein Bound** → Part X, Section 1.4 (Information Capacity)
- **Quantum Entanglement** → Part X, Section 1.2 (Entanglement)
- **Emergence** → Part X, Section 1.5 (concepts of emergence in consciousness chapter)

LOOKING AHEAD:

- Chapter 6 examines metaphysical questions about what it means "to be"
- Chapter 7 explores consciousness as another emergent but irreducible phenomenon
- Part III synthesizes how holographic/informational structure connects to divine creation

- The relational nature revealed here prefigures Trinitarian relationality in Part V

TECHNICAL REFERENCE:

- For clarification that holography is mathematical description, not metaphysical claim: Part X, Section 1.3 (Holographic Cosmology - Caution)
- For what "emergence" means and doesn't mean: Part X, Section 7.4 (Principle of Non-Reduction)
- For observer-independence in quantum mechanics: Part X, Section 1.2 (Measurement Problem)

NOTE:

All references to "observers" in physics contexts throughout this book refer to physical systems capable of interaction that produces definite outcomes—not conscious minds. This distinction is explained in detail in Part X, Section 6.. The claim that "information structures the universe" is mathematical, not ontological—information-theoretic language describes physical processes without implying that information is conscious, purposeful, or more fundamental than matter and energy.

---

1. Bekenstein, J. D. (1973). "Black Holes and Entropy." *Physical Review D* 7(8), 2333–2346. Bekenstein proposed that black holes must possess entropy proportional to their horizon area to preserve the second law of thermodynamics.
2. Hawking, S. W. (1975). "Particle Creation by Black Holes." *Communications in Mathematical Physics* 43(3), 199–220. Hawking demonstrated that black holes emit thermal radiation and calculated the precise relationship between entropy and horizon area: $S = A/4$ in Planck units.
3. 't Hooft, G. (1993). "Dimensional Reduction in Quantum Gravity." arXiv:gr-qc/9310026. Susskind, L. (1995). "The World as a Hologram." *Journal of Mathematical Physics* 36(11), 6377–6396. These papers formulated the holographic principle, proposing that the information content of a spatial region is encoded on its boundary.
4. Maldacena, J. (1998). "The Large N Limit of Superconformal Field Theories and Supergravity." *Advances in Theoretical and Mathematical Physics* 2(2), 231–252. The AdS/CFT correspondence provides the most concrete realization of holography, establishing an exact equivalence between gravitational theories in Anti-de Sitter space and conformal field theories on the boundary.
5. Ryu, S., & Takayanagi, T. (2006). "Holographic Derivation of Entanglement Entropy from AdS/CFT." *Physical Review Letters* 96(18), 181602. This formula

relates entanglement entropy in the boundary theory to the area of minimal surfaces in the bulk spacetime, providing a geometric interpretation of quantum entanglement.
6. Bekenstein, J. D. (1981). "Universal Upper Bound on the Entropy-to-Energy Ratio for Bounded Systems." *Physical Review D* 23(2), 287–298. The Bekenstein bound establishes the maximum entropy (information content) that can be contained within a given region with a given energy.
7. Hawking, S. W. (1974). "Black Hole Explosions?" *Nature* 248(5443), 30–31. This brief paper introduced the concept of Hawking radiation and black hole evaporation, leading to the information paradox that continues to drive research in quantum gravity.

CHAPTER FOUR
TIME, ENTROPY, AND THE ARROW OF BECOMING

Time is everywhere and nowhere. You experience it constantly with things like aging, remembering, and anticipating. And yet you cannot touch it, measure it directly, or slow it down. Clocks tick forward. Milk spoils. Stars burn out. Children grow up. The universe moves relentlessly in one direction, and we call that direction the future. This feels so natural that we rarely question it. But physics reveals something startling: the fundamental laws governing reality do not care which way time flows.

Write down any of the equations that describe motion. Newton's laws, Maxwell's electromagnetism, Einstein's relativity, even the Schrödinger equation of quantum mechanics. They work perfectly well whether time runs forward or backward. Mathematically, past and future are symmetric. A film of billiard balls colliding could be reversed and still obey all the laws of physics. The same holds for planetary orbits, electromagnetic waves, and quantum evolution. Nothing in the equations themselves demands a preferred direction. Yet the universe clearly has one.

We age. We do not grow younger. Stars exhaust their fuel. They do not gather it back. Coffee cools to room temperature. It does not spontaneously heat up. Ice melts. Disorder spreads. The world moves from order toward chaos with absolute consistency. This directionality is so deeply embedded in our experience that we take it for granted. But it is not explained by the fundamental laws alone. Something else accounts for time's arrow. That something is entropy.

Entropy is often described as disorder, but that description can mislead. More precisely, entropy measures the number of microscopic arrangements, or microstates, that correspond to a given macroscopic condition. A system with low entropy has very few possible configurations that match its observable state. A system with high entropy has vastly more.

Consider a deck of cards arranged by suit and rank. This ordered state represents low entropy because only a handful of arrangements match the description "sorted perfectly." Now shuffle the deck. After shuffling, the cards are scrambled. This disordered state represents high entropy because countless arrangements fit the description "randomly shuffled." The transition from order to disorder is not magic or even surprising. It is statistical inevitability. There are simply more ways for the deck to be disordered than ordered.

The same principle governs the universe. The early cosmos existed in a state of extraordinarily low entropy. Despite being unimaginably hot and dense, the distribution of matter and energy was remarkably smooth and uniform. Gravitational degrees of freedom were barely activated. The universe was simple, symmetric, exquisitely ordered. This is what cosmologists mean when they say the universe began in a low-entropy state.

As the universe expanded, that initial simplicity gave way to increasing complexity. Gravity pulled matter into clumps. Stars formed from collapsing gas clouds. Galaxies assembled from billions of stars. Planets condensed from stellar debris. At every stage, entropy increased—not because chaos spread but because the universe explored more and more of the vast space of possible configurations. The low-entropy beginning was improbable. The progression toward higher entropy is the universe moving from the unlikely toward the overwhelmingly probable.

This increase in entropy is not merely a trend. It is a law. The second law of thermodynamics states that in any closed system, entropy never decreases over time.[1] It either increases or remains constant. This law is statistical rather than absolute—technically, entropy could decrease in a tiny region for a brief moment—but for any macroscopic system over any significant duration, the second law holds with near certainty. And it is this law that gives time its arrow.

The arrow of time is the arrow of increasing entropy. Time flows forward because entropy grows. Every irreversible process—burning fuel, breaking an egg, aging—corresponds to an increase in entropy. The reason you remember the past but not the future is that the past represents a state of lower entropy, and your memories are records formed by entropy-increasing processes. The reason causes always precede effects is that the low-entropy beginning provides directionality to causation. The universe started ordered and has been moving toward greater entropy ever since.

But this raises a profound puzzle. Why did the universe begin in such an extraordinarily low-entropy state? Why was the early cosmos smooth, uniform, and simple rather than chaotic and maximally disordered? If entropy naturally increases, and if the universe had existed eternally, it should have reached maximum entropy long ago. Yet it did not. It began in a condition of remarkable order, and that initial condition is what makes time's arrow possible.

Cosmologists have no complete explanation for this low-entropy beginning. Some speculate about inflationary mechanisms that might have set the initial conditions. Others explore quantum cosmology or multiverse scenarios. But these remain speculative.[2] What is clear is that the low-entropy beginning is not a prediction of physics. It is an observed feature of the universe that physics presupposes but does not explain. This is a boundary where physics reaches its limit and deeper questions emerge: Why this particular initial state? Why order rather than chaos? Why a universe capable of giving rise to structure, life, and consciousness?

The relationship between entropy and time also sheds light on the nature of structure. At first glance, the claim that entropy always increases might seem to contradict the evident growth of order in the universe. Stars form. Galaxies organize. Life emerges. Complex systems develop. Is this not a decrease in entropy?

Not quite. The formation of structure involves a local decrease in entropy, but it is always accompanied by a larger increase elsewhere. When a gas cloud collapses to form a star, the star represents a local concentration of order. But the collapse releases tremendous energy into the surrounding space. That energy disperses, carrying entropy with it. The net result is an overall increase in entropy. The universe as a whole becomes more disordered even as pockets of order emerge within it.

Life itself is a spectacular example of this principle. Living organisms are highly ordered systems. They maintain low internal entropy by continually processing energy from their environment. They eat, breathe, metabolize, and excrete. In doing so, they export entropy to their surroundings. A living organism creates more disorder in the universe than the order it sustains within itself. Life does not violate the second law of thermodynamics. It exploits it.[3]

This interplay between local order and global disorder is essential to understanding cosmic evolution. The universe moves from simplicity toward complexity not by reversing entropy but by exploring the vast space of possible configurations. Structure forms because gravity, nuclear forces, and quantum mechanics allow certain stable arrangements to emerge. These arrangements persist because they are stable, not because they decrease total entropy. The universe becomes more interesting over time precisely because it is moving toward higher entropy, not despite it.

Quantum mechanics adds another layer to this story. At the quantum level, systems exist in superpositions—states where multiple possibilities coexist simultaneously until an interaction occurs. When a quantum system interacts with its environment, the superposition collapses and one outcome becomes definite. (Note carefully: "collapse" here refers to a physical interaction producing a definite outcome, not observation by a conscious mind—this distinction is explained fully in Part X, Section 6.) This process, called decoherence, is not a violation of quantum mechanics. It is a consequence of entanglement with the environment.[4]

Decoherence plays a crucial role in the emergence of the classical world from the quantum substrate. It is the process by which quantum potentialities give way to definite outcomes. And every time a quantum system decoheres, information about the outcome becomes encoded in the environment. This encoding is an entropy-increasing process. It is irreversible. Once the environment has recorded the outcome, there is no returning to the original superposition. The quantum state has become definite not because a conscious observer looked but because the system interacted with the world in a way that left a permanent trace.

This connection between quantum decoherence and entropy reinforces the arrow of time. The direction of time is the direction in which quantum systems decohere, in which possibilities become actualities, and in which distinctions are recorded. Time flows forward because the universe is constantly selecting outcomes from quantum potentials and encoding those outcomes into its structure. The past is fixed because it has been recorded. The future is open because it has not. In thermodynamic terms, the past leaves records—traces in memory, matter, and radiation. The future has not yet left such records.

Physicist Rolf Landauer discovered another profound connection between information and thermodynamics. His principle states that erasing information has a thermodynamic cost.[5] To erase a single bit of information, or to reset a memory element from one or zero back to a neutral state, requires a minimum amount of energy proportional to the temperature of the system. This means that information is physical. It is not an abstract concept floating above the material world. It is embedded in the physical states of systems, and manipulating it requires work.

Landauer's principle has far-reaching implications. It shows that every act of computation, every measurement, every recording of a distinction comes with an energy cost. The universe generates information continually as it evolves, and each new piece of information contributes to the overall increase in entropy. (Again, "information" here is mathematical, not conscious—see Part X, Section 6 for this critical distinction.) The generation of information and the increase of entropy are two descriptions of the same underlying process: the universe exploring the space of possible configurations.

The arrow of time, then, is not an illusion or a human projection. It is a feature of the universe grounded in the laws of thermodynamics, the dynamics of quantum systems, and the statistical behavior of complex systems. Time has a direction because the universe began in an improbable, low-entropy state and has been evolving toward higher entropy ever since. This directionality makes causation possible, memory possible, and meaningful change possible. Without the arrow of time, there would be no history, no growth, no development. There would only be static existence or endless reversible cycles.

Standing here at the intersection of physics and philosophy, we encounter questions that physics alone cannot answer. Physics describes how entropy increases and how the arrow of time emerges from thermodynamic principles. But it does not explain why the universe began in a low-entropy state. It does not tell us why there is becoming at all, why change is real, or why the universe moves from simplicity toward complexity rather than remaining in eternal equilibrium.

These questions point beyond mechanism to meaning. They ask not merely how time works but why there is time. They ask not merely what entropy is but what it signifies. They are metaphysical questions, and they demand metaphysical answers.

From a theological perspective, the arrow of time resonates deeply with the biblical vision of creation. Genesis does not describe a static cosmos but a dynamic unfolding. Creation progresses from formlessness to order, from void to fullness, from evening to morning. Each day brings something new. The biblical narrative is fundamentally historical—it tells a story with direction, purpose, and culmination. Time is not cyclical, endlessly repeating. It is linear, moving toward fulfillment.

The God revealed in Scripture is not timeless in the sense of being frozen or inert. God acts, speaks, creates, and relates. The divine name "I AM" does not signify static being but the ground of all becoming, the source from which change and history flow. The universe has an arrow of time because it is the creation of a God who works purposefully through history toward redemption and consummation.

This does not mean Genesis predicts entropy or that thermodynamics proves Scripture. It means that the directionality revealed by physics harmonizes with the directionality described in theology. Both point to a universe that is not eternal and static but temporal and dynamic. Both describe a reality with a beginning, a history, and a trajectory. The arrow of time is not an accident. It is a feature of a created order designed to unfold, develop, and move toward an end.

The next section of this book turns explicitly to philosophy. Having traced the physical structure of reality—its beginning, its informational character, its holographic depth, and its temporal direction—we now ask what that structure means. What does it mean for anything to exist? What is consciousness, and why does it emerge in a universe governed by physical law? What is freedom, and what does it reveal about the nature of persons? These questions build upon the insights of physics, but they require a different mode of inquiry. The journey from mechanism to meaning continues in Chapter 5.

CHAPTER REVIEW:
KEY CONCEPTS:

- **Fundamental physics laws are time-symmetric, yet we experience directional time** - Newton's laws, Maxwell's equations, even Schrödinger equation work equally well forward or backward

- **The arrow of time emerges from thermodynamics** - Entropy (disorder) increases because the universe began in an extraordinarily low-entropy state
- **Time is the dimension along which potentiality becomes actuality** - The universe moves from simple initial conditions toward increasing complexity and information
- **Decoherence explains classical behavior from quantum substrates** - Environmental interaction produces definite outcomes without requiring conscious observation

WHAT THIS MEANS:

The directionality of time - past, present, future - is not fundamental to physics equations but emerges from initial conditions and statistical mechanics. The fact that the universe had such special initial conditions (low entropy) is itself unexplained by physics and points toward metaphysical questions about why the universe exists in this particular state.

IMPORTANT TERMS & CONCEPTS:

- **Entropy** (thermodynamic) → Part X, Section 1.3 (Second Law of Thermodynamics)
- **Arrow of Time** → Part X, Section 1.3 (Time - thermodynamic)
- **Decoherence** → Part X, Section 1.2 (Decoherence)
- **Measurement** (quantum) → Part X, Section 1.2 (Measurement Problem)
- **Actualization** (potency to act) → See Chapter 6; Part X, Section 1.6 (Act vs. Potency)

LOOKING AHEAD:

- Chapter 5 addresses why the arrow of time points beyond physics to metaphysics
- Chapter 6 develops the metaphysical concepts of potency and act crucial for understanding becoming
- Chapter 11 shows creation as God actualizing potential through time
- Part IV connects the progressive unfolding of time to Genesis and Logos theology

TECHNICAL REFERENCE:

- For clarification that observers need not be conscious: Part X, Section 1.2 (Observer/Measurement Problem)
- For the relationship between information and entropy: Part X, Section 1.4 (Entropy-Information Relationship)
- For what "becoming" means philosophically vs. physically: Part X, Section 1.6 (Act vs. Potency; Actualization)

NOTE:

Throughout this chapter, terms like "measurement" and "collapse" refer to physical interactions producing definite outcomes, not to conscious observation. The universe's arrow of time emerges from thermodynamic and quantum processes, not from human awareness. These clarifications are detailed in Part X.

---

1. Clausius, R. (1865). "Über verschiedene für die Anwendung bequeme Formen der Hauptgleichungen der mechanischen Wärmetheorie." *Annalen der Physik und Chemie* 125, 353–400. Clausius formulated the second law of thermodynamics and introduced the concept of entropy, establishing that in any isolated system, entropy never decreases.
2. For discussions of the low-entropy initial conditions and attempts to explain them through inflation and quantum cosmology, see: Penrose, R. (1989). "Difficulties with Inflationary Cosmology." *Annals of the New York Academy of Sciences* 571, 249–264; Carroll, S. M., & Chen, J. (2004). "Spontaneous Inflation and the Origin of the Arrow of Time." arXiv:hep-th/0410270. These remain active areas of research without consensus.
3. Schrödinger, E. (1944). *What Is Life? The Physical Aspect of the Living Cell.* Cambridge University Press. Schrödinger's classic work explores how living organisms maintain order by exporting entropy, a foundational insight in biophysics.
4. Zurek, W. H. (2003). "Decoherence, Einselection, and the Quantum Origins of the Classical." *Reviews of Modern Physics* 75(3), 715–775. Decoherence explains how quantum superpositions give way to definite classical states through interaction with the environment, without requiring conscious observation.
5. Landauer, R. (1961). "Irreversibility and Heat Generation in the Computing Process." *IBM Journal of Research and Development* 5(3), 183–191. Landauer's principle establishes the fundamental thermodynamic cost of information erasure, linking computation to the second law of thermodynamics.

PART TWO
THE PHILOSOPHY BENEATH THE PHYSICS

CHAPTER FIVE
WHERE PHYSICS STOPS AND PHILOSOPHY BEGINS

You've followed the journey this far. Through four chapters, we've traced what modern physics reveals about the universe: a cosmos with a beginning, structured by information, governed by mathematical elegance, and directed by the arrow of time. These discoveries represent some of humanity's greatest intellectual achievements. They are genuine, empirically grounded, and scientifically uncontroversial.

But standing here, having surveyed what physics shows us, something unexpected emerges. The more clearly we see what physics reveals, the more clearly we see what it cannot reveal. Not because physics has failed—it hasn't. We've simply reached the boundary where one kind of inquiry ends and another must begin.

This boundary is not arbitrary. It's built into the nature of scientific investigation itself. Physics excels at answering certain questions with extraordinary precision. But there are other questions—equally important, equally legitimate, equally real—that belong to a different domain entirely. These aren't questions about mechanisms or measurements. They're questions about existence, meaning, consciousness, and purpose. They're philosophical questions. And they require philosophical answers.

Understanding where this boundary lies matters more than most people realize. In our time, confusion about these limits has created false conflicts and unnecessary divisions. Some claim science can answer every meaningful question. Others retreat from scientific findings as if they threaten deeper truths. Both mistakes arise from the same confusion: failing to recognize that different questions require different modes of inquiry.

This chapter marks the transition from physics to philosophy. It doesn't abandon what we've learned, it builds upon it. It takes seriously what physics reveals and asks what those revelations mean. The journey from mechanism to meaning begins here.

Consider the most fundamental question we can ask: Why does anything exist at all?

Physics describes the behavior of matter and energy across cosmic history with breathtaking precision. It traces the universe's evolution from the Big Bang forward, showing how structure emerged from simplicity, how galaxies formed, how stars ignited, how planets condensed. This story is magnificent in its scope and detail. But it presupposes existence. It begins with something already there: energy, spacetime, quantum fields, the laws governing their behavior.

The Big Bang marks the universe's temporal beginning, the point where our current physics breaks down and the classical concepts of space and time cease to apply. But a temporal beginning is not the same as an explanation for existence itself. To say the universe began expanding from an initial state doesn't answer why there was an initial state. It doesn't explain why there is something rather than nothing.

That question stands behind physics, prior to any equation or measurement. It's not a scientific question because science studies what exists, not why existence occurs at all. This isn't a gap in current knowledge that future discoveries might fill. It's a question that falls outside the scope of what science investigates. Recognizing this isn't a criticism of physics, it's clarity about what physics actually is and what it was designed to do.

This realization opens onto deeper territory. If physics cannot explain existence, what can? The question shifts from mechanism to ground, from description to foundation. It becomes a question about being itself, what philosophers call ontology. And here we encounter a profound insight that has shaped philosophical thought for millennia: everything in the physical universe is contingent.

Nothing we observe contains within itself the reason for its own existence. Stars depend on nuclear forces. Atoms depend on quantum fields. Planets depend on gravitational attraction. Even the laws of nature themselves could have been otherwise. The speed of light could have been different. The strength of gravity could have varied. The electromagnetic force could have taken another value. Everything points beyond itself to something it depends upon.[1]

This contingency creates what philosophers recognize as an explanatory regress. If everything depends on something else, either the chain extends infinitely, or it terminates in something that doesn't depend on anything further. An infinite regress explains nothing, it merely pushes the question back endlessly without ever providing a ground.[2] Therefore, reason demands a terminus: something whose essence is existence itself, something that cannot not be, something upon which all contingent things depend but which depends on nothing.

Classical metaphysics, from Aristotle through Aquinas to contemporary analytic philosophers, identifies this necessary ground with what we call God.[3] But we're getting ahead of ourselves. For now, the point is simpler: physics reveals contingency everywhere. Philosophy asks what contingency requires. And the answer points toward a ground of being that physics, by its nature, cannot investigate.

The same pattern appears when we ask why the universe is intelligible. Science depends entirely on this intelligibility. Experiments yield consistent results. Patterns repeat. Regularities can be expressed mathematically. Laws discovered in one context apply universally. The behavior of atoms on Earth matches the behavior of atoms in distant galaxies.

But why? Why should nature conform to elegant mathematical structures? Why should the equations we derive from observing falling apples also describe the motion of planets? The physicist Eugene Wigner captured this puzzle perfectly when he called it "the unreasonable effectiveness of mathematics in the natural sciences."[4] It's unreasonable because there's no obvious necessity for it. Mathematics arises from human thought, yet it describes the cosmos with stunning precision.

Physics uses mathematics. It doesn't explain why mathematics works. That explanatory gap points toward something profound: the universe is structured in a way that rational minds can grasp. There's a deep correspondence between thought and reality, between the structure of our reasoning and the structure of nature. This correspondence isn't itself a physical fact, you can't measure it or put it in an equation. It's a condition that makes physics possible.

And recognizing it as a condition rather than a conclusion opens philosophical questions about the relationship between mind and world, between intelligibility and existence, between the knower and the known. These are questions about the rational structure of reality itself, questions that point toward a rational ground of being.

Then there's consciousness. Modern neuroscience has mapped brain activity with remarkable detail. We can identify which regions activate during perception, memory, decision-making, and emotion. We can predict certain behaviors based on neural patterns. We can even stimulate specific brain areas to produce particular experiences.

But none of this explains why there is experience at all. Why does it feel like something to be conscious? Why is there an interior dimension to reality, a first-person perspective that cannot be captured by any third-person description? You know what it's like to see red, to feel pain, to experience joy. But no amount of information about neurons firing tells us what it is like to have these experiences.

Philosophers call this the hard problem of consciousness, and it resists every attempt at physical reduction.[5] The qualities of experience, what philosophers call qualia, are irreducibly subjective. They belong to the conscious subject and cannot be observed from outside. You can scan my brain while I look at a sunset, but you'll never see the sunset as I experience it. You'll only see neural correlations with that experience.

This isn't a claim that consciousness violates physics or exists in some supernatural realm separate from the body. The relationship between mind and brain is intimate and undeniable. But the relationship itself, the fact that physical processes give rise to subjective experience, remains philosophically perplexing. Physics describes the mechanisms. Philosophy asks what those mechanisms mean and why they produce an inward life at all.

Consider one more boundary: morality. We all operate with moral intuitions. Torturing children for fun is wrong. Keeping promises matters. Treating people with equal dignity is important. Some actions are genuinely better than others, not just useful or pleasant, but actually right.

Yet where do these truths come from? Evolution can explain why we have moral instincts, cooperation aids survival. But evolution explains why we believe certain things, not whether those beliefs are true. The fact that kindness conferred evolutionary advantage doesn't tell us whether kindness is actually good. It only tells us why we think it is.

Physics can describe the neural correlates of moral judgment. It can map which brain regions activate when we make ethical decisions. But physics cannot tell us whether any moral judgment is correct. It cannot derive "ought" from "is." The gap between description and prescription, between what happens and what should happen, cannot be crossed by empirical observation alone.[6]

Biology might explain why humans evolved to value their offspring. But it doesn't explain why all human beings possess equal dignity regardless of age, ability, or achievement. That belief, foundational to Western civilization and enshrined in documents like the Declaration of Independence, is not a biological observation. It's a metaphysical commitment, historically grounded in theological claims about human nature bearing the image of God.

What these examples reveal is a pattern. Physics, philosophy, and theology are not competing explanations for the same phenomena. They are distinct modes of inquiry addressing fundamentally different questions. Understanding this distinction is crucial, not just for intellectual clarity but for avoiding the false conflicts that have plagued discussions of science and faith for generations.

Physics describes mechanisms. It answers questions about how things work: How does gravity operate? What are the laws governing quantum systems? How did the universe evolve from its initial state? Physics is empirical, mathematical, and predictive. It deals with measurable quantities, testable hypotheses, and natural processes. Its domain is the behavior of matter and energy within spacetime.

Metaphysics (philosophy) explains being. It answers questions about what it means to exist, what grounds contingent reality, why the universe is intelligible, and what consciousness is. Philosophy is rational, conceptual, and foundational. It examines the principles that underlie all experience and knowledge. Its domain is the nature of existence itself.

Theology addresses ultimate cause and purpose. It answers questions about why there is a universe at all, what the source of being is, whether reality has meaning, and what our ultimate destiny is. Theology is revelational (drawing on Scripture) while also engaging philosophically. It deals with the personal ground of all beings, God, and the relationship between Creator and creation. Its domain is the transcendent source and final end of all things.

These domains are not isolated. They inform one another. Physics provides data that philosophy interprets. Philosophy clarifies concepts that theology employs. Theology offers a framework within which both physics and philosophy find their place. But they must not be confused. Each operates with its own methods, answers its own questions, and maintains its own standards of evidence and argument.

When we forget these distinctions, confusion follows. People treat physics as if it could answer metaphysical questions, then claim science has disproven God. Or they treat theology as if it were a competing scientific theory, then reject biblical truth because it doesn't make empirical predictions. Or they demand that philosophy adopt the methods of physics, then dismiss all non-empirical reasoning as meaningless.

These are category errors: treating one kind of question as if it belonged to another domain. And they're as absurd as asking what color justice is, or how much a musical note weighs, or where happiness is located in space. The question itself reveals a misunderstanding about the nature of the thing being asked about.

Let me set this in stone. Physics tells us that the universe began in a low-entropy state approximately 13.8 billion years ago. That's a description of what happened. It's empirically well-supported and scientifically uncontroversial. But it doesn't explain why the universe began in that particular state. It doesn't tell us why there's a universe capable of having a low-entropy beginning. Those are philosophical questions.

Physics tells us the fundamental constants are fine-tuned for the existence of complexity. That's an observation. It's well-documented and precisely measured. But it doesn't tell us why the constants have the values they do. It doesn't explain whether this fine-tuning is the result of necessity, chance, or design. Those are metaphysical and theological questions.

Physics tells us consciousness correlates with brain activity. That's a fact. It's demonstrable and measurable. But it doesn't explain why physical processes produce subjective experience. It doesn't tell us what consciousness is or why it exists at all. Those are philosophical questions that require philosophical methods to address.

The point is not that physics is limited or inadequate. The point is that physics is what it is, a spectacularly successful enterprise for understanding the behavior of nature, and we shouldn't ask it to be something else. Expecting physics to explain existence is like expecting a hammer to function as a screwdriver. It's not a failing of the hammer. It's a misunderstanding of what hammers are for.

You might wonder why this matters. Why spend a chapter clarifying boundaries between disciplines? Because the alternative is intellectual chaos.

If you don't understand where physics stops, you'll either expect it to answer every question (and despair when it can't) or reject its findings because they seem incomplete (and miss the genuine insights it provides). If you don't understand what philosophy contributes, you'll dismiss metaphysical reasoning as mere speculation (and cut yourself off from understanding the foundations of your own thought). If you don't understand theology's proper role, you'll either treat it as bad science (and reject it for the wrong reasons) or treat science as a threat to faith (and create unnecessary conflict).

The synthesis this book offers depends entirely on respecting these boundaries. Physics shows us a universe that is rational, relational, information-theoretic, and finely structured. Philosophy asks what kind of reality could produce such a universe and arrives at the need for a necessary ground of being that is itself rational and relational. Theology identifies that ground with the God revealed in Scripture. The eternal I AM, Being Itself, the Logos through whom all things were made.

These are not three different answers to the same question. They are complementary perspectives on a single reality, each operating within its proper domain, each illuminating truths the others cannot access alone. When properly understood, they converge on a coherent vision. But only when we maintain the distinctions that allow each to function as it should.

The chapters ahead will build on this foundation. We'll explore what it means for anything to exist, how consciousness fits into the structure of reality, what freedom reveals about personhood, and why the physical universe points toward a personal, relational, transcendent ground. But every step depends on the clarity we've established here: physics, philosophy, and theology are distinct. Each has its role. Each deserves respect. And together, they reveal more than any could alone.

The next chapter turns to the deepest of metaphysical questions: What does it mean for something to exist? What is being itself? And how does the answer to that question illuminate everything we've learned about the universe physics reveals?

CHAPTER REVIEW:

KEY CONCEPTS

- **Physics describes mechanisms; philosophy explains being** - Science tells us HOW things work, but cannot answer WHY anything exists at all

- **Domain boundaries must be respected** - Physics, philosophy (metaphysics), and theology address fundamentally different questions using different methods

- **Existence, intelligibility, consciousness, and morality cannot be reduced to physics** - These are legitimate questions requiring philosophical/metaphysical answers

- **Category errors occur when questions are asked in the wrong domain** - Asking "what color is justice?" reveals confusion about the nature of the thing in question

WHAT THIS MEANS:

There are natural limits to what physics can explain - not because science has failed, but because certain questions fall outside its scope by definition. Recognizing these boundaries prevents false conflicts between science and faith and clarifies what each discipline actually claims. Both are necessary for a complete understanding of reality.

IMPORTANT TERMS & CONCEPTS:

- **Physics** (domain of) → Part X, Section 1.1 (Domain Boundaries: Physics)

- **Metaphysics** (domain of) → Part X, Section 1.1 (Domain Boundaries: Metaphysics)
- **Theology** (domain of) → Part X, Section 1.1 (Domain Boundaries: Theology)
- **Category Error** → Part X, Section 2 (entire section on Category Errors)
- **Contingency** → Part X, Section 1.6 (Contingent Being)

LOOKING AHEAD:
- Chapter 6 develops the metaphysical framework needed to understand existence itself
- All subsequent chapters operate with awareness of these domain boundaries
- Part X provides the comprehensive technical treatment of domain separation
- Part III (Chapter 10) shows how respecting boundaries enables genuine synthesis

TECHNICAL REFERENCE:
- For comprehensive domain boundary clarifications: Part X, Section 1 (entire section)
- For examples of category errors to avoid: Part X, Section 2
- For what synthesis does and doesn't mean: Part X, Section 7 (Integration Guidelines)
- For common misunderstandings about science/faith relationship: Part X, Section 4

NOTE ON DOMAIN BOUNDARIES:

This chapter establishes the foundational distinction between physics, philosophy, and theology that structures the rest of this book. Subsequent chapters will reference these domains using shorthand ("As established in Chapter 5, these domains remain distinct..."). For comprehensive treatment of domain boundaries, category errors, and methodological clarifications, see Part X, Sections 1–2. The principle is simple: physics describes mechanisms, metaphysics explains being, theology addresses ultimate cause. Respecting these boundaries prevents confusion and allows genuine synthesis.

1. The contingency of physical systems and natural laws is a central theme in classical metaphysics. For contemporary treatments, see: Koons, R. C., & Pickavance, T. H. (2017). *The Atlas of Reality: A Comprehensive Guide to Metaphysics*. Wiley-Blackwell. For historical development: Aquinas, T. *Summa Theologica*, Part I, Questions 2–3.
2. The problem of infinite regress in causal explanations has been addressed by philosophers from Aristotle onward. See: Aristotle. *Metaphysics*, Book XII (Lambda). For modern formulations: Pruss, A. R. (2006). *The Principle of Sufficient Reason: A Reassessment*. Cambridge University Press.
3. Classical arguments from contingency to a necessary being include Aquinas's Third Way and Leibniz's argument from sufficient reason. For rigorous contemporary defenses: Rasmussen, J., & Leon, F. (2018). *Is God the Best Explanation of Things? A Dialogue*. Palgrave Macmillan. Feser, E. (2017). *Five Proofs of the Existence of God*. Ignatius Press.
4. Wigner, E. (1960). "The Unreasonable Effectiveness of Mathematics in the Natural Sciences." *Communications in Pure and Applied Mathematics* 13(1), 1–14. Wigner's essay remains the definitive statement of this profound puzzle about the relationship between mathematics and physical reality.
5. Chalmers, D. J. (1996). *The Conscious Mind: In Search of a Fundamental Theory*. Oxford University Press. Chalmers's work distinguishes the "easy problems" of consciousness (explaining cognitive functions) from the "hard problem" (explaining why there is subjective experience at all). For further discussion: Nagel, T. (1974). "What Is It Like to Be a Bat?" *The Philosophical Review* 83(4), 435–450.
6. The is-ought problem, first articulated clearly by David Hume, remains central to moral philosophy. See: Hume, D. (1739). *A Treatise of Human Nature*, Book III, Part I, Section I. For contemporary engagement: Pigliucci, M., & Boudry, M. (Eds.). (2013). *Philosophy of Pseudoscience: Reconsidering the Demarcation Problem*. University of Chicago Press.

CHAPTER SIX
WHAT IT MEANS TO EXIST: THE METAPHYSICS OF BEING

*U*p to this point, you've explored the physical structure of the universe: its beginning, its relational foundation, the mathematics that shape it, the arrow of time that unfolds through increasing entropy, and the emergence of distinct structures. Yet physics, by its nature, describes only **how** the universe behaves. It does not, and cannot, address the deeper question that stands behind all scientific inquiry: why does anything exist at all?

This chapter introduces a crucial step, one that many books skip completely: the metaphysical bridge. Thomas Aquinas faced this same transition 800 years ago. He has Aristotle's natural philosophy (the physics of his day). He had Christian theology. And he recognized that moving directly from one to the other without philosophical groundwork would be intellectually irresponsible.

So he put in the hard work that is required of Metaphysics—not to 'prove' God with science, but to show how the natural world (contingent, changing, ordered) could cohere with the God of classical theism (necessary, eternal, simple).

The same work Aquinas did centuries ago applies to modern cosmology today. Its not speculation. Its not unnecessary. Its the bridge that makes genuine synthesis possible. Stay with it. The intellectual tools you gain here will serve you for the rest of the book, and far beyond it.

While physics presupposes existence, law, intelligibility, and relational order—it cannot fully explain them. Metaphysics seeks to understand what these features mean. And in doing so, it opens the door to insights that later chapters will connect to Scripture, classical theology, and the divine name "I AM."

The first question is simple to ask but difficult to answer: what does it mean for something to exist? In the physical sciences, existence is usually defined operationally. A thing exists if it has causal effects, if it can be measured or detected. But this definition is circular. Measurement presupposes the existence of the measuring system. Detection presupposes a world in which detection is possible. Physics offers no account of why anything is detectable in the first place.

Metaphysics approaches existence differently. It asks what it means for a thing to "be", independent of any observational framework. Classical metaphysics, from Plato to Aquinas, distinguishes between essence and existence. Essence is "what" a thing is. Existence is "that" a thing is. This distinction may seem abstract, but it cuts to the core of reality. A triangle has an essence defined by its properties, yet the concept of a triangle does not guarantee that any actual triangle exists. Being is something added to essence.[1]

Modern physics, though it uses different language, reflects this distinction. Quantum mechanics describes the possible states a system may occupy, the essence, but those possibilities do not become actual until an interaction occurs. Possibility alone does not generate actuality. There must be a transition, a movement from what may be to what is. This movement is what physicists call collapse or decoherence, but philosophically, it is the transition from potency to act, a distinction thoroughly mapped by classical metaphysics.

The universe itself, understood through cosmology, exhibits this same pattern. The early universe contained a range of potentialities encoded in quantum fluctuations. Over time, these potentialities unfolded into structure. Galaxies, stars, planets, living organisms, and conscious beings represent the progressive actualization of possibilities embedded in the initial conditions. The universe is a history of becoming, a continuous movement from potential to actuality.

This raises a profound question: why does the universe move from potency to act at all? Why is there becoming? Why is there change? Classical metaphysics holds that a system cannot actualize itself. Something already in act must actualize what is in potency. This insight points toward the need for a grounding act of being, a source that is not merely another contingent part of the universe but the very condition for the existence of contingent things.[2]

Physics encounters a similar limit when confronting the origin of the universe. The Big Bang cannot be explained as the result of earlier physical causes. Time and causality themselves emerged from it. The beginning of the universe points toward something beyond physical explanation, not because science has failed, but because the origin of time cannot be contained within temporal categories. The universe depends on something that does not itself depend on space, time, or matter.

Another aspect of existence that metaphysics clarifies is intelligibility. The universe is not chaotic. It is ordered according to laws that can be expressed in mathematical form. This mathematical structure is not imposed by human observers. It is woven into the fabric of reality. The question is why. Why is the universe intelligible? Why does it obey stable, rational laws? Why can human minds grasp these laws?

Physics describes the laws. Metaphysics asks why laws exist and why they are rational. Classical metaphysics answers this by arguing that intelligibility reflects being. What exists is structured, ordered, and capable of being known because it participates in the rational order of being itself. The universe is intelligible because being is intelligible.[3] Later chapters will show how this insight resonates deeply with the doctrines of the Logos and the divine name.

Another metaphysical feature of existence is relationality. Things do not exist in isolation. They exist through relationships. Quantum entanglement shows that even at the smallest scales, particles are defined by their relations to one another. Space itself emerges from relational patterns of entanglement. Identity is relational, not isolated. This relational structure mirrors metaphysical principles articulated throughout philosophical and theological history: that being is inherently relational, that to be is to stand in relation.[4]

Information theory strengthens this observation. Information is defined by distinctions. A bit is meaningful only relative to other bits. A system is defined by the differences that make a difference. Existence is therefore tied to relational differentiation. Something exists because it is distinct from what it is not. The universe exists as a web of relations, a tapestry of complexity unfolding across time.

This relational nature of being connects directly to the ancient theological insight that God is relational within Himself and relational toward creation. But that argument belongs to later chapters. For now, it is enough to recognize that the universe operates according to principles of relational being. Existence is not solitary. It is relational at its core.

The final metaphysical feature to consider is contingency. Everything we observe in the universe could have been otherwise. The laws of physics could have been different. The constants of nature could have taken other values. The distribution of matter could have varied. Even the existence of the universe itself is not metaphysically necessary. Nothing about the physical world contains the reason for its own existence. It depends on something beyond itself.

Classical metaphysics refers to this dependency as contingency. A contingent being is one whose essence does not entail its existence. The universe, with its finely tuned laws and initial low-entropy state, is a profoundly contingent system. It points beyond itself to a necessary ground of being, something whose essence is existence, something that cannot not exist.[5] Later chapters will connect this insight with the meaning of the divine name.

This chapter has shown that the universe's physical features—its relationality, its intelligibility, its movement from potentiality to actuality, its contingency—cannot be fully explained by physics alone. As established in Chapter 5, physics describes the structure of becoming. Metaphysics explains what becoming means. The two are complementary, not competitive. Each operates within its proper domain, and together they illuminate truths that neither could access alone.

The next chapter will turn to the emergence of consciousness, the inward dimension of reality. Consciousness is the point where the universe becomes aware of itself, the moment where relational structures take on an interior life. It is here that the metaphysical and the experiential meet, setting the stage for understanding personhood, freedom, and the unique role of human beings within the unfolding universe.

CHAPTER REVIEW:

KEY CONCEPTS:

- **Essence vs. Existence** - What a thing IS (essence) is distinct from THAT it is (existence); existence is added to essence

- **Potency vs. Act** - Things contain unrealized possibilities (potency) that become actual through actualization

- **lNothing can actualize itself** - Something already in act must actualize what is in potency; this points toward a necessary ground of being

- **Contingent beings require a necessary being** - Everything in the universe COULD have been otherwise; none contain the reason for their own existence

- **Being itself is relational** - Quantum mechanics reveals that particles are defined by relationships, not isolated properties

WHAT THIS MEANS:

Classical metaphysics provides the conceptual tools needed to understand what existence itself requires. The universe's contingency - the fact that it need not exist - points beyond itself to a necessary ground. This is not a "God of the gaps" argument but recognition that physics presupposes existence without explaining it. Metaphysics addresses this deeper question.

IMPORTANT TERMS & CONCEPTS:
- **Essence** → Part X, Section 1.6 (Essence)
- **Existence** → Part X, Section 1.6 (Existence as actuality)
- **Act and Potency** → Part X, Section 1.6 (Act vs. Potency; Actualization)
- **Contingent Being** → Part X, Section 1.6 (Contingent Being)
- **Necessary Being** → Part X, Section 1.6 (Necessary Being)
- **Substance** → Part X, Section 1.6 (Substance)

LOOKING AHEAD:
- Chapter 10 connects these concepts to God as "I AM" (Being Itself)
- Chapter 11 shows creation as actualization of potential
- Chapter 16 develops "I AM" as the metaphysical name revealing God's nature as pure act
- Part V explores how being itself is relational, not isolated

TECHNICAL REFERENCE:
- For complete metaphysical terminology: Part X, Section 1.6 (entire subsection)
- For how metaphysics relates to physics: Part X, Section 7.10 (Ontological Priority)
- For causality (four causes): Part X, Section 1.6 (Causality - Four Causes)
- For participation in being: Part X, Section 1.6 (Participation)

---

1. The distinction between essence and existence is central to classical metaphysics, particularly in the work of Thomas Aquinas. For a rigorous contemporary treatment, see: Gilson, É. (1952). *Being and Some Philosophers*. Pontifical Institute of Mediaeval Studies. For Aquinas's original formulation: Aquinas, T. *Summa Theologica*, Part I, Question 3, Article 4. The principle that existence is an act added to essence establishes that finite beings are composites, participating in being rather than possessing it by nature.
2. The principle that potency cannot actualize itself without something already in act is a foundational insight of Aristotelian-Thomistic metaphysics. See: Aristotle. *Metaphysics*, Book XII (Lambda), especially Chapters 6–7. For contemporary

philosophical defense: Feser, E. (2017). *Five Proofs of the Existence of God*. Ignatius Press, pp. 19–66. This principle underlies the cosmological argument and explains why every change points toward an unchanging source.

3. The intelligibility of being is explored extensively in classical realism. For the relationship between being and knowability, see: Lonergan, B. (1992). *Insight: A Study of Human Understanding*. University of Toronto Press. For the claim that intelligibility is intrinsic to being itself: Gilson, É. (1952). *Being and Some Philosophers*, Chapter 4. The correspondence between mind and reality is not arbitrary but reflects the rational structure of existence.

4. The relational nature of being has roots in both philosophical and theological traditions. For philosophical treatments, see: Norris Clarke, W. (2001). *The One and the Many: A Contemporary Thomistic Metaphysics*. University of Notre Dame Press, especially Chapter 4 on "Person and Being." For theological perspectives on relationality as fundamental to being, see: Zizioulas, J. D. (1985). *Being as Communion: Studies in Personhood and the Church*. St Vladimir's Seminary Press.

5. The concept of contingency and the argument from contingency to a necessary being has a long philosophical history. For classical formulations, see: Aquinas, T. *Summa Theologica*, Part I, Question 2, Article 3 (the Third Way). Leibniz, G. W. (1697). "On the Ultimate Origination of Things." For contemporary rigorous defenses: Pruss, A. R. (2006). *The Principle of Sufficient Reason: A Reassessment*. Cambridge University Press. Rasmussen, J., & Leon, F. (2018). *Is God the Best Explanation of Things? A Dialogue*. Palgrave Macmillan. The contingency argument shows that whatever exists but need not exist requires an explanation beyond itself, ultimately pointing toward a being whose very essence is existence.

CHAPTER SEVEN
CONSCIOUSNESS: THE UNIVERSE WAKING UP TO ITSELF

Consciousness is the most intimate certainty of human existence and the greatest challenge to any purely physical account of the world. It is the single phenomenon that each person encounters directly, yet the one that physicists and philosophers struggle most to explain. You know consciousness from the inside —as awareness, experience, perception, memory, emotion, and meaning. But what is it in the structure of reality? How does the universe give rise to beings that experience themselves as subjects rather than objects?

This chapter explores consciousness in the context of a relational universe. It does not attempt to solve the hard problem of consciousness, nor does it reduce consciousness to physics. Instead, it examines how consciousness fits into a universe governed by information, how it emerges in the unfolding of complexity, and why it occupies a central place in any coherent metaphysics of being.

The first step is recognizing that consciousness is not an incidental byproduct of matter. It is not a chemical accident. It is a fundamental aspect of what it means for a universe to have interiority. Every human experience begins with the simple fact that we are aware. We have thoughts. We perceive the world. We reflect on ourselves. This interior life is not something physics can describe in quantitative terms, but it is real. Consciousness is not visible to measurement, yet it is more certain than any measurement could ever be.

In a relational universe, consciousness can be viewed as the integration of information into a unified perspective. Modern cognitive science refers to this as integrated information: the synthesis of sensory input, memory, reasoning, and internal states into a coherent whole.[1] A conscious being has an interior model of the world and of itself. It is able to reflect on its own existence, to imagine, to choose, and to recognize meaning.

This integration requires complexity, but not just any complexity. It requires structure that can hold together multiple streams of information and bind them into a single point of view. Biological organisms have evolved systems capable of this integration. The human brain is the most complex known structure in the observable universe, containing billions of neurons interconnected in patterns that can represent the world, respond to it, and reflect upon it.

Physics does not yet explain how consciousness arises from this structure, but it does tell us something essential: consciousness depends on information processing, and information processing is woven into the fabric of the universe. At the quantum level, systems hold multiple possibilities until interactions generate new information. At larger scales, entanglement structures the geometry of space. Within living systems, information is copied, transmitted, and transformed through biochemical and neural networks.

Consciousness is the point at which information becomes experience. It is where the universe moves from the exterior world of facts and relations into the interior world of meaning. The relational fabric of reality provides the conditions under which consciousness can emerge, but consciousness itself introduces a fundamentally new dimension: subjectivity.

Subjectivity cannot be captured by any third-person description. A neuroscientist can measure brain activity, but no amount of external measurement can reveal what it is like to experience a thought or a feeling. This interior dimension is not reducible to matter or energy. It reflects a mode of being that is irreducibly first-person.[2]

The existence of subjectivity raises another profound question: why does the universe allow for beings that experience themselves from the inside? Why does the universe unfold in such a way that consciousness appears? The relational structure of reality provides one part of the answer. As the universe evolves, it generates increasingly complex structures capable of integrating information. Stars form. Planets emerge. Life adapts. Nervous systems evolve. Eventually, consciousness appears as a natural extension of the universe's relational trajectory.

But this explanation addresses only the "how," not the "why." The deeper question is why the universe has an interior dimension at all. Why should information give rise to experience? Why should something in the universe know that it exists?

Classical metaphysics offers a perspective that resonates with these questions. According to classical thought, being is not merely existence. Being includes awareness. To be is to be intelligible. To be intelligible is to be ordered toward mind. Consciousness is not a foreign element in the universe. It is the highest expression of what it means for being to be knowable.[3]

This perspective is strengthened by the relational structure of reality uncovered by quantum mechanics and information theory. Consciousness is the integration of relationships into a coherent whole. It is the point at which relational structure becomes interior life. In this sense, consciousness is not something that stands apart from the universe. It is the universe becoming aware of itself through beings capable of reflection.

This idea does not imply that the universe is conscious in a simple or literal way. It does imply that consciousness fits naturally into a world whose foundation is relational and informational. The universe contains the seeds of consciousness because it contains the structure necessary for awareness. Consciousness is the highest form of relational integration, the most refined expression of informational unity.

Human consciousness introduces another dimension: self-awareness. You not only experience the world; you experience yourself experiencing the world. This recursive awareness allows for reasoning, morality, creativity, and the capacity to choose. It allows you to ask questions about your origin, your purpose, and your destiny. It opens the possibility of relationship with that which transcends the physical world.

The mystery of consciousness also points toward the limits of a purely physical explanation. If consciousness cannot be reduced to physical processes, then it suggests that reality has a dimension beyond the physical—a dimension that aligns with the metaphysical and theological insights of later chapters. Consciousness is the bridge between physics and personhood, between the world of information and the world of meaning.

The final aspect to consider is that consciousness is not isolated. It is inherently relational. Human beings develop self-understanding through interaction with others. Knowledge grows through dialogue. Meaning arises through relationship. This relational nature of consciousness mirrors the relational structure of reality itself. As we established earlier, the universe is not a collection of isolated objects. It is a network of relationships. Consciousness is the personal expression of that relational order.[4]

This chapter has shown that consciousness fits naturally into a relational universe. It arises through the integration of information, expresses the interior dimension of being, and reveals the relational structure of reality. It also raises deep questions about the ultimate source of consciousness, questions that lead beyond physics into the realm of the personal and the divine.

The next chapter will turn to freedom, agency, and the emergence of persons. If consciousness is the inward dimension of the universe, then personhood is its highest relational form. Understanding what it means to be a person will lead us directly into the

Consciousness: The Universe Waking Up to Itself

questions of dignity, morality, and the image of the One who is "I AM."

CHAPTER REVIEW:

KEY CONCEPTS:

- **Consciousness is subjective, first-person experience (qualia)** - There is "something it is like" to be conscious that cannot be captured by third-person description
- **The hard problem: why does physical processing produce subjective experience?** - No amount of describing brain states explains WHY there is inner experience
- **Consciousness cannot be reduced to computation or neural activity** - Correlation between brain and mind does not equal identity
- **Intentionality (aboutness) characterizes mental states** - Thoughts are ABOUT things; physical states just are
- **Unity of consciousness resists reductionist explanation** - The integration of diverse experiences into one coherent field requires more than neural synchronization

WHAT THIS MEANS:

Consciousness reveals a dimension of reality that cannot be captured by physics alone. This doesn't require rejecting neuroscience - brain states clearly correlate with conscious states - but it recognizes that subjective experience is irreducible. Consciousness points toward something beyond mechanism, opening questions about the nature of persons and their place in reality.

IMPORTANT TERMS & CONCEPTS:

- **Consciousness** → Part X, Section 1.5 (Consciousness - General Definition)
- **Qualia** → Part X, Section 1.5 (Qualia)
- **Intentionality** → Part X, Section 1.5 (Intentionality)
- **Subjectivity** → Part X, Section 1.5 (Subjectivity)
- **Hard Problem** → Part X, Section 1.5 (The Hard Problem)

- **Unity of Consciousness** → Part X, Section 1.5 (Unity of Consciousness)

LOOKING AHEAD:
- Chapter 8 builds on this to explore freedom and moral agency
- Chapter 22 examines personal identity and being known by I AM
- Part V develops how persons image the relational God
- The irreducibility of consciousness parallels the irreducibility of personhood

TECHNICAL REFERENCE:
- For complete consciousness terminology: Part X, Section 1.5 (entire subsection)
- For why consciousness ≠ information: Part X, Section 7.4 (Principle of Non-Reduction)
- For mind vs. brain distinction: Part X, Section 1.5 (Mind - Philosophical Definition)
- For common errors about consciousness: Part X, Section 4 (Misunderstandings)

---

1. Integrated Information Theory (IIT), developed by Giulio Tononi, proposes that consciousness corresponds to integrated information in a system. See: Tononi, G. (2004). "An Information Integration Theory of Consciousness." *BMC Neuroscience* 5, 42. While IIT offers a mathematical framework for quantifying consciousness, it does not resolve the hard problem of why physical integration produces subjective experience. The theory is controversial but provides a useful framework for thinking about consciousness in information-theoretic terms. For critical discussion: Koch, C., Massimini, M., Boly, M., & Tononi, G. (2016). "Neural Correlates of Consciousness: Progress and Problems." *Nature Reviews Neuroscience* 17(5), 307–321.
2. The distinction between first-person and third-person perspectives is central to the philosophy of mind. For the definitive statement of the hard problem, see: Chalmers, D. J. (1996). *The Conscious Mind: In Search of a Fundamental Theory*. Oxford University Press. Chalmers distinguishes the "easy problems" of consciousness (explaining cognitive functions like perception and memory) from the "hard problem" (explaining why there is subjective experience at all). For the classic thought experiment about subjective experience: Nagel, T. (1974). "What Is It Like to Be a Bat?" *The Philosophical Review* 83(4), 435–450. Nagel argues that the subjective character of experience cannot be captured by objective, physical descriptions.

3. The classical view that being is inherently intelligible and ordered toward mind has roots in Aristotelian and Scholastic philosophy. For Aristotle's account of intellect and intelligibility, see: Aristotle. *De Anima* (On the Soul), Book III. For Aquinas's development of this theme: Aquinas, T. *Summa Theologica*, Part I, Questions 14–16. For a contemporary treatment connecting intelligibility to consciousness: Lonergan, B. (1992). *Insight: A Study of Human Understanding*. University of Toronto Press. The principle that mind and being are fundamentally related suggests consciousness is not accidental but intrinsic to the rational structure of reality.

4. The relational nature of consciousness is explored in phenomenology and personalist philosophy. For the development of self-consciousness through social interaction, see: Mead, G. H. (1934). *Mind, Self, and Society*. University of Chicago Press. For philosophical treatments of consciousness as inherently relational: Buber, M. (1937/1970). *I and Thou*. Translated by Walter Kaufmann. Scribner. For theological perspectives on relationality and personhood: Zizioulas, J. D. (1985). *Being as Communion: Studies in Personhood and the Church*. St Vladimir's Seminary Press. These works establish that consciousness develops through relationship and reflects the relational structure of being itself.

CHAPTER EIGHT
FREEDOM, AGENCY, AND THE EMERGENCE OF PERSONS

*I*f consciousness is the inward dimension of the universe, then freedom and agency are its outward expression. Consciousness allows you to experience the world from the inside; agency allows you to act within it. Personhood, in turn, arises when consciousness and agency converge in a relational, self-aware subject capable of knowledge, intention, and moral responsibility. This chapter examines the emergence of persons in a universe grounded in information and relational structure.

Freedom and agency are often treated as philosophical or psychological topics, but they also have a place in the scientific story of the universe. The universe began in simplicity and moved toward complexity. Structure emerged from the interplay of gravitational and quantum processes. As complexity increased, systems gained the ability to regulate themselves, respond to their environments, and eventually initiate actions. Agency did not appear suddenly. It emerged along a continuum.

At its most basic level, agency exists wherever a system can respond to information. Even simple organisms sense nutrients, avoid danger, and adapt to changing conditions. These responses are not conscious in the human sense, yet they demonstrate an elementary form of directed behavior. As biological systems became more complex, they developed internal models of their surroundings. Nervous systems appeared. Learning became possible. Memory allowed organisms to carry the past into the present. Agency increased.

Human beings represent the highest known form of agency. You do not merely react to stimuli; you deliberate. You weigh options. You imagine possible futures. You choose based on values, goals, and understanding. Your freedom is not absolute, but it is real. Human agency is an expression of personhood, the capacity to act in ways that reflect rationality, intention, and moral discernment.

What, then, is the metaphysical basis of freedom? In a deterministic universe governed by fixed laws, freedom seems impossible. But the universe described in earlier chapters is not deterministic in that sense. Quantum mechanics introduces genuine possibilities into the fabric of reality. Information theory shows that the universe generates new distinctions as it unfolds. The relational structure of the universe allows for openness, emergence, and novelty. The universe is not a closed mechanical system. It is a dynamic, unfolding reality.[1]

Classical metaphysics provides another layer of insight. Freedom is not the absence of constraints but the presence of the capacity to act according to one's nature. A stone has no freedom because it lacks an interior principle of action. A plant has limited freedom because it responds automatically to environmental cues. An animal has more freedom because it possesses sensation, desire, and mobility. A rational being has the highest degree of freedom because it possesses intellect and will. The human person can understand, deliberate, and choose.[2]

This hierarchy is not arbitrary. It reflects the structure of being. To exist as a person is to exist with the capacity for rational agency. This understanding aligns with the relational universe described earlier. A person integrates information, forms intentions, and acts in ways that shape the world. A person participates in the relational structure of reality not only as a receiver but as a contributor.

Freedom also introduces moral responsibility. To act freely is to bear responsibility for those actions. This responsibility presupposes that you could have acted differently. It presupposes awareness of value, the ability to recognize right and wrong, and the capacity to direct your actions toward the good. Moral responsibility reflects the personal dimension of the universe. It cannot be reduced to physics. It belongs to the realm of persons.

The emergence of moral agency raises another question that physics alone cannot answer: why does the universe produce moral beings at all? Why does the universe give rise to creatures capable of knowing the good, seeking justice, valuing truth, and loving others? These capacities are not reducible to survival mechanisms. They reflect a dimension of reality that transcends mere physical processes.[3]

This transcendence points toward a deeper metaphysical ground. If the universe is intelligible, relational, and oriented toward meaning, then the emergence of persons who recognize meaning is not an accident. It is an expression of the universe's underlying structure. Persons occupy a unique place in the universe. We are not isolated observers but participants in the universe's unfolding relational story. Our choices alter the world. Our understanding adds to the universe's intelligibility. Our relationships mirror the relational structure of reality itself.

Personhood also involves selfhood. A person is a self, a center of consciousness and action that persists through time. This persistence cannot be explained solely by physical continuity. The body changes constantly, yet personal identity endures. Memory, intention, and self-awareness knit together the continuity of the self. This unity is not reducible to matter. It reflects a metaphysical principle: the person as a substantial being whose identity is grounded in more than physical processes.[4]

Relationality deepens this understanding. Persons become

themselves through relationships. A child develops self-understanding through interaction with parents, peers, and the world. Adults grow in knowledge and virtue through dialogue, community, and love. Personhood flourishes in relationship because persons are inherently relational. This relational structure echoes the metaphysical insight that being itself is relational.

Freedom, agency, selfhood, and relationality converge in the concept of dignity. A person possesses inherent worth not because of abilities or attributes but because of the kind of being the person is. Dignity is rooted in the nature of personhood. It cannot be taken away by circumstances, lost by weakness, or diminished by failure. It is fundamental.[5] This understanding of dignity will play a crucial role in later chapters, especially in the discussion of natural law, human rights, and the foundations of liberty.

This chapter has shown that persons emerge naturally within a relational universe. Consciousness provides the inward dimension. Freedom provides the outward dimension. Agency expresses the unity of rational intention and action. Selfhood grounds personal identity. Relationality reflects the structure of being. Dignity arises from the nature of personhood itself.

The next chapter turns to a historical example that powerfully illustrates the synthesis of scientific inquiry and theological devotion. Isaac Newton, widely considered one of the greatest minds in human history, devoted more of his intellectual energy to understanding God than to understanding physics. His life demonstrates that the pursuit of truth in science and the pursuit of truth in theology are not competitors but complementary perspectives on a single reality.

CHAPTER REVIEW:
KEY CONCEPTS:

- **Freedom is the capacity to act according to one's nature** - Not absence of constraints but presence of rational agency

- **Moral responsibility presupposes genuine freedom** - If you could not have acted otherwise, responsibility makes no sense

- **Persons occupy a unique place in the universe** - Integration of rationality, relationality, consciousness, and agency

- **The universe's relational structure permits openness and emergence** - Quantum indeterminacy and biological variability allow for genuine novelty

- **Hierarchy of freedom corresponds to hierarchy of being** - Stones < plants < animals < rational persons in capacity for self-directed action

WHAT THIS MEANS:

The emergence of free, rational agents is not accidental but reflects the universe's deep structure. Freedom requires both the right kind of physical order (laws stable enough to support agency) and metaphysical openness (genuine possibilities). The fact that the universe produces moral beings capable of recognizing value points toward questions physics cannot answer.

IMPORTANT TERMS & CONCEPTS:

- **Agency** → Part X, Section 1.5 (concepts of free will and personhood)

- **Freedom** (philosophical) → Part X, Section 1.6 (concepts of act and self-determination)

- **Person** → Part X, Section 1.6 (Subsistence; concepts of persons)

- **Moral Responsibility** → See Chapter 20; Part X treatment of moral law

- **Emergence** → Part X, Section 7.4 (Principle of Non-Reduction)

LOOKING AHEAD:

- Chapter 12 explores why genuine freedom requires the possibility of imperfection

- Chapter 19 shows how Christianity uniquely grounds human dignity and freedom

- Chapter 22 develops personal identity before I AM

- Part VII examines living as free persons in relation to God

TECHNICAL REFERENCE:

- For hierarchy of being: Part X, Section 1.6 (Ontological Hierarchy)

Freedom, Agency, and the Emergence of Persons

- For freedom and divine sovereignty: Part X, Section 5 (Anticipated Objections)
- For teleology without determinism: Part X, Section 1.6 (Finality - Teleology)
- For personhood as relational: See also Part V (Trinity chapters)

---

1. The question of determinism and freedom in light of modern physics is a long-standing philosophical issue. Quantum indeterminacy introduces genuine randomness at the microscopic level, though whether this grounds human freedom remains debated. For discussion of quantum mechanics and determinism, see: Kane, R. (Ed.). (2002). *The Oxford Handbook of Free Will*. Oxford University Press, especially chapters on physics and free will. For the emergence of novelty in complex systems: Prigogine, I., & Stengers, I. (1984). *Order Out of Chaos: Man's New Dialogue with Nature*. Bantam Books. These works explore how openness in physical systems creates space for emergent properties including agency.

2. The classical distinction between levels of being and corresponding levels of freedom derives from Aristotelian and Scholastic philosophy. For Aristotle's account of different souls (vegetative, sensitive, rational) and their capacities, see: Aristotle. *De Anima* (On the Soul), Books II–III. For Aquinas's development of this hierarchy: Aquinas, T. *Summa Theologica*, Part I, Questions 75–79. For contemporary philosophical treatment: Feser, E. (2014). *Scholastic Metaphysics: A Contemporary Introduction*. Editiones Scholasticae. These sources establish that freedom increases with the complexity and interiority of a being's nature.

3. The emergence of moral agency and its resistance to purely evolutionary or physical explanation is explored in contemporary philosophy of mind and ethics. For arguments that morality transcends survival mechanisms, see: Nagel, T. (2012). *Mind and Cosmos: Why the Materialist Neo-Darwinian Conception of Nature Is Almost Certainly False*. Oxford University Press. For the relationship between rationality and moral insight: Plantinga, A. (2011). *Where the Conflict Really Lies: Science, Religion, and Naturalism*. Oxford University Press, Chapter 9. These works argue that moral knowledge points toward a deeper rational order in reality.

4. The persistence of personal identity through change is a classic problem in philosophy. For the substance view of persons versus competing theories, see: Swinburne, R. (2013). *Mind, Brain, and Free Will*. Oxford University Press, especially Chapters 7–8 on personal identity. For the role of memory and psychological continuity: Parfit, D. (1984). *Reasons and Persons*. Oxford University Press, Part III. For a hylomorphic (form-matter) account of personal identity: Stump, E. (2003). *Aquinas*. Routledge, Chapter 6. These discussions clarify why personal identity requires more than physical continuity.

5. The concept of intrinsic human dignity has deep roots in Christian theology and classical philosophy. For the imago Dei (image of God) as the foundation of dignity, see: Genesis 1:26–27. For theological development: Aquinas, T. *Summa*

Theologica, Part I, Question 93. For contemporary philosophical defense of inherent dignity: Spaemann, R. (2006). *Persons: The Difference Between 'Someone' and 'Something'*. Translated by Oliver O'Donovan. Oxford University Press. For dignity's role in ethics and politics: George, R. P. (1999). *In Defense of Natural Law*. Oxford University Press, Chapter 6. These sources establish that dignity is not conferred by society or earned through achievement but inherent in personhood itself.

CHAPTER NINE
ISAAC NEWTON AND HIS TRUE DEVOTION

*I*saac Newton is widely considered one of the most intelligent human beings who ever lived. His *Philosophiæ Naturalis Principia Mathematica* laid the foundation for classical mechanics. His work on calculus revolutionized mathematics. His experiments with light transformed our understanding of optics. His law of universal gravitation explained the motion of planets and apples alike. He didn't just advance science, he invented entire fields of scientific inquiry.

Newton asked how the universe works. And he answered that question with unprecedented precision.

But Newton also asked why the universe exists, who made the laws he discovered, and what it all means. And he devoted more of his intellectual energy to these questions than he did to physics. Newton wrote more than a million words on theology and biblical interpretation.[1] He studied prophecy, church history, and the nature of God with the same rigor he applied to planetary motion. To him, investigating the natural world was a way of thinking God's thoughts after Him. Science and theology were not competitors. They were complementary perspectives on a single reality.

Newton didn't think you had to choose between being scientifically literate and theologically serious. He pursued both with equal passion. And he is not an exception. Many of the founders of modern science were deeply religious.

Galileo Galilei, despite his famous conflict with Church authorities, remained devoutly Catholic his entire life. Even as he fought for the freedom to pursue astronomical truth, he saw no contradiction between his scientific work and his faith. He famously wrote that Scripture tells us "how to go to heaven, not how the heavens go."[2] His conflict was with ecclesiastical authorities about scriptural interpretation, not with faith itself.

Johannes Kepler saw his astronomical work as revealing the mind of God. When he discovered the laws of planetary motion, he wrote: "I was merely thinking God's thoughts after him. Since we astronomers are priests of the highest God in regard to the book of nature, it benefits us to be thoughtful, not of the glory of our minds, but rather, above all else, of the glory of God."[3] For Kepler, every equation was an act of worship.

Gregor Mendel, the father of genetics, was an Augustinian monk. His groundbreaking experiments with pea plants were conducted in the monastery garden. He saw no tension between his religious vocation and his scientific curiosity. Both were ways of understanding God's creation.[4]

Georges Lemaître, who first proposed what we now call the Big Bang theory, was a Catholic priest. When he presented his theory of the "primeval atom" in 1927, showing that the universe had a beginning, some scientists resisted precisely because it sounded too much like Genesis. But Lemaître carefully maintained the distinction between physics and theology. He understood that science describes how the universe works, while theology addresses why it exists and what it means.[5]

Robert Boyle, one of the founders of modern chemistry, wrote extensively about how scientific investigation leads naturally to contemplation of God's wisdom.

Francis Collins, who led the Human Genome Project, is an evangelical Christian who sees his work as "a form of worship."[6]

The list goes on: Blaise Pascal, Michael Faraday, James Clerk Maxwell, William Thomson (Lord Kelvin)—brilliant minds who saw no contradiction between rigorous science and deep faith.

The idea that science and faith are inherently opposed is not a conclusion these thinkers reached. It is a recent cultural construction that distorts history and impoverishes thought. So why does this false choice persist? Why are you told today that you must pick a side?

Part of the answer lies in the way the question "why?" has been split in two. There are actually two very different kinds of "why," and confusion between them has caused enormous damage.

The first is the scientific why. This is the question about mechanisms and causes. Why does the apple fall? Gravity. Why do species change over time? Evolution. Why does the universe expand? The dynamics described by general relativity. Scientific "why" asks how things work. It seeks explanations in terms of laws, forces, interactions, and processes. Science is brilliantly suited to answering this kind of question.

The second is the philosophical why. This is the question about meaning, purpose, and existence itself. Why does anything exist at all? Why is the universe intelligible? Why do I have the capacity to ask questions? Why does my life matter? Philosophical "why" asks what it all means. It seeks understanding in terms of value, purpose, identity, and the good. Science was never designed to answer this kind of question.

These two questions are not in competition. They are complementary. But somewhere along the way, we started treating them as if they were the same question with competing answers. And that confusion created a false binary: you're either scientifically literate or you care about meaning. You're rational or you're religious. You trust evidence or you have faith.

This is nonsense.

Newton understood that both questions are real, both are important, and both deserve serious answers. The apple falls because of gravity. And gravity exists because God established the laws of nature. Evolution explains how species adapt. And the fact that the universe produces conscious beings capable of discovering evolution points toward something deeper. The Big Bang describes cosmic origins. And the question of why there is a cosmos at all remains unanswered by physics.

The people who built the civilization you inherited didn't believe it was either/or. The American Founders were brilliant Enlightenment rationalists fluent in multiple languages, steeped in classical learning, and conversant with cutting-edge science. Yet they grounded their entire political philosophy in theological commitments. When the Declaration of Independence states that rights are "endowed by their Creator," that is not decoration. It is the foundation. Rights don't come from government because they come from something higher than government. Human dignity isn't a social construct because humans bear the image of God.[7]

Remove the theology and you get a different politics. Not because the Founders were ignorant, but because they understood something that seems to have been lost today: some truths require both scientific rigor and philosophical depth and theological insight. The universe is a physical system governed by laws. And those laws point toward a rational source. And that source is personal, relational, and worthy of worship.

Newton knew this. Kepler knew this. Lemaître knew this. The greatest minds in history have known this. They didn't think asking both kinds of "why" made them less intelligent. They thought it made them more fully human.

You've probably experienced this yourself: asking "why?" until the answers run out. Why is the sky blue? Why do birds fly? Physics has answers for these. Why does anything happen at all? Why does physics work? Some questions point beyond science, to something deeper. Perhaps to something your parents called "Just because."

At the time, you may have thought they were dodging the question. But "just because" is actually a profound answer. It acknowledges that at some point, explanation reaches bedrock. Not every "why" has a further "why" behind it. At some point, you encounter the brute fact that existence simply is. And you're here, conscious, wondering about it.

Just because.

That phrase captures something essential. It's the recognition that there are limits to explanation, but those limits don't mean the questions stop mattering. In fact, the questions become more important. When science reaches the boundary of what it can explain, philosophy and theology take over. Not as competitors to science, but as the natural continuation of the same inquiry.

The universe is intelligible. That's a fact of physics. Why is it intelligible? That's a question for metaphysics. The universe follows elegant mathematical laws. That's an empirical observation. Why do those laws exist? That requires philosophy. The universe produces conscious beings who can understand it. That's undeniable. What does that reveal about the nature of reality? That's where theology enters.

These aren't different conversations. They're chapters in the same story. And the story makes sense only when you read all the chapters.

The remaining chapters of this book follow in Newton's footsteps. You'll explore the synthesis of physics, philosophy, and theology—not by forcing them together, but by recognizing that they illuminate different aspects of the same reality. You'll see how modern cosmology points toward questions that physics cannot answer. You'll examine how classical metaphysics provides the conceptual tools needed to understand being itself. And you'll discover how Christian theology reveals the personal God who is the ground of all existence.

This is not an exercise in apologetics that tries to "prove" God from equations. It's an invitation to see coherence where you've been taught to see conflict. It's a recognition that the smartest people in history didn't abandon either reason or faith, They held both, and in holding both, they changed the world.

The next section turns to the (IAM) framework itself. The synthesis that shows how physics, philosophy, and theology converge when each operates within its proper domain. You'll see why the universe described by modern science is precisely the kind of universe we would expect if it were grounded in the God who reveals Himself as "I AM."

CHAPTER REVIEW:

KEY CONCEPTS:

- **Newton wrote over a million words on theology** - Far exceeding his scientific output; his faith was central, not peripheral
- **The two kinds of "why"** - Scientific why (mechanism) vs. philosophical why (meaning/purpose)
- **History's greatest scientists were often deeply religious** - Galileo, Kepler, Mendel, Lemaître, Boyle, Pascal, Maxwell, Faraday, Collins
- **The false choice between science and faith is culturally constructed** - Not a conclusion reached by the founders of modern science
- **"Just because" points to bedrock** - Some explanations reach foundations where further "why" requires different modes of inquiry

WHAT THIS MEANS:

The conflict narrative between science and faith is historically inaccurate. The people who built modern science saw no contradiction between rigorous empirical investigation and deep theological commitment. Both ask legitimate questions; both deserve serious answers. The choice is not either/or but recognizing which questions belong to which domain.

IMPORTANT TERMS & CONCEPTS:

- **Two kinds of "Why"** Scientific mechanism vs. philosophical purpose (see Chapter 5)
- **Natural Theology** → Historical tradition of reasoning about God from nature
- **Domain Boundaries** → Part X, Section 1 (entire section)
- **Galileo's dictum** Scripture tells "how to go to heaven, not how the heavens go"

LOOKING AHEAD:

- Part III (Chapters 10-12) presents the synthesis Newton and others exemplified
- Chapter 19 traces how Christian theology shaped Western intellectual tradition
- The rest of the book demonstrates that holding both perspectives enriches understanding
- This chapter serves as transition from physics/philosophy to theological synthesis

TECHNICAL REFERENCE:
- For historical examples of science-faith harmony: See endnotes in this chapter
- For methodological principles allowing synthesis: Part X, Section 7 (Integration Guidelines)
- For what synthesis does NOT claim: Part X, Section 3
- For avoiding category errors in interdisciplinary work: Part X, Section 2

---

1. Newton's theological writings comprise over one million words, far exceeding his scientific output. For comprehensive treatment of Newton's theology, see: Snobelen, S. D. (1999). "Isaac Newton, Heretic: The Strategies of a Nicodemite." *The British Journal for the History of Science* 32(4), 381–419. For Newton's own theological manuscripts: Newton, I. *The Chronology of Ancient Kingdoms Amended* (1728); *Observations upon the Prophecies of Daniel and the Apocalypse of St. John* (1733). Newton's theological concerns were central to his life's work, not peripheral to it.
2. This famous formulation appears in Galileo's "Letter to the Grand Duchess Christina" (1615), where he argues for the independence of scriptural interpretation from empirical science. For full text and context: Galilei, G. (1615/1957). "Letter to the Grand Duchess Christina." In *Discoveries and Opinions of Galileo*, translated by Stillman Drake. Doubleday Anchor, pp. 173–216. Galileo's conflict was with Aristotelian natural philosophy conflated with theology, not with Christianity itself.
3. Kepler, J. (1619). *Harmonices Mundi* (The Harmony of the World), Book V, Chapter 9. For English translation and commentary: Kepler, J. (1997). *The Harmony of the World*, translated by E. J. Aiton, A. M. Duncan, and J. V. Field. American Philosophical Society. Kepler explicitly saw his astronomical work as a priestly vocation, revealing the mathematical order God built into creation.
4. For Mendel's life as monk and scientist, see: Orel, V. (1996). *Gregor Mendel: The First Geneticist*. Oxford University Press. Henig, R. M. (2000). *The Monk in the Garden: The Lost and Found Genius of Gregor Mendel, the Father of Genetics*.

Houghton Mifflin. Mendel's abbot encouraged his scientific work, and the monastery provided resources for his experiments.

5. For Lemaître's development of Big Bang cosmology and his careful distinction between physics and theology, see: Lemaître, G. (1927). "Un Univers homogène de masse constante et de rayon croissant rendant compte de la vitesse radiale des nébuleuses extra-galactiques." *Annales de la Société Scientifique de Bruxelles* A47, 49–59. For biographical context: Lambert, D. (2015). *The Atom of the Universe: The Life and Work of Georges Lemaître*. Copernicus Center Press. Lemaître insisted that his physics was independent of his theology, though both informed his worldview.

6. Collins, F. S. (2006). *The Language of God: A Scientist Presents Evidence for Belief*. Free Press, p. 233. Collins describes how reading C.S. Lewis led him from atheism to Christianity while maintaining his commitment to rigorous science. His work leading the Human Genome Project demonstrates no conflict between cutting-edge biology and Christian faith.

7. For the theological foundations of the American founding and natural rights theory, see: Dreisbach, D. L., & Hall, M. D. (Eds.). (2014). *The Sacred Rights of Conscience: Selected Readings on Religious Liberty and Church-State Relations in the American Founding*. Liberty Fund. West, T. G. (1997). "The Political Theory of the Declaration of Independence." In *The American Founding: Essays on the Formation of the Constitution*, edited by J. J. Basile and F. Canavan. University Press of Kansas. The Declaration's appeal to "Nature's God" and "Creator" reflects Enlightenment natural theology grounded in classical Christian metaphysics.

PART THREE
(IAM) THE INFORMATIONAL ACTUALIZATION MODEL

CHAPTER TEN
DEFINING THE INFORMATIONAL ACTUALIZATION MODEL

*Y*ou've come a long way on this journey. You've explored the universe as physics reveals it. A universe beginning in a Big Bang, structured by information, governed by elegant laws, and unfolding through time's arrow. You've examined where physics reaches its natural limits and where philosophical questions emerge: Why does anything exist? Why is the universe intelligible? What is consciousness? You've seen that these questions require answers beyond the reach of empirical science.

Now you arrive at the central claim of this book. The Informational Actualization Model is not a scientific theory competing with established physics. It is not a new equation that predicts supernovae or galaxy rotation curves. It is something different: a synthesis showing how modern physics, classical philosophy, and Christian theology converge on a coherent vision of reality. Each discipline asks different questions and uses different methods. But when properly understood, with each operating within its own domain, can illuminate a single unified picture.

This chapter defines what the (IAM) actually is, what it claims, and what it does not claim. Clarity here matters because confusion about these boundaries has created unnecessary conflict between science and faith. The goal is not to collapse physics into theology or to derive biblical doctrine from equations. The goal is to show that what physics describes, what philosophy requires, and what theology reveals are three perspectives on one reality.

The (IAM) proposes that the universe described by modern physics as rational, relational, informational, and intelligible; is precisely the kind of universe we would expect if it were grounded in the God revealed in Christian Scripture: the eternal (IAM), Being Itself, the rational Logos, the relational Trinity.

The name itself carries meaning. (IAM) stands for **Informational Actualization Model**, a deliberate echo of God's self-revelation in Exodus 3:14 (ESV): "God said to Moses, 'I AM WHO I AM.'" But this is not a physics theory competing with general relativity or quantum mechanics. It is a theological and philosophical framework for understanding how creation unfolds.[1]

Think of it this way: The universe grows in informational content; think complexity, structure, and meaningful distinctions, as it moves from the "formless void" toward ever-greater order. This is actualization in the classical metaphysical sense: potency becoming act, possibility becoming reality. The framework describes what Scripture teaches and what physics observes: reality coming into being through divine speech, order emerging through meaningful distinction, potentials being actualized according to divine purpose.

This is not an attempt to prove God's existence through scientific observation. Physics cannot prove metaphysical claims. But it can reveal patterns that resonate with theological truths. And when those patterns align with insights from philosophy and theology, the coherence is worth noting.

Imagine you're examining a magnificent tapestry. One person studies it with a magnifying glass, counting threads, analyzing dye composition, measuring weave patterns. Another steps back and considers the overall design, asking what principles of artistry and composition make the image work. A third person knows the artist personally and understands the story the tapestry tells, the meaning it conveys, and why it was created.

All three perspectives are true. None contradicts the others. The thread-counter isn't wrong about fiber composition. The design analyst isn't wrong about compositional principles. The friend of the artist isn't wrong about meaning and purpose. Together, they provide a richer understanding than any single view could offer.

That's what the (IAM) proposes for reality itself.

Physics is the magnifying glass. It maps the structure with extraordinary precision. The universe is not a static container of isolated objects. It is a dynamic web of relationships. Particles do not exist independently; they are defined by their interactions. Space itself emerges from relational patterns of quantum entanglement. Information is woven into the fabric of reality. The laws of nature are elegant, mathematically expressible, and universal. The cosmos began in a low-entropy state and has been generating structure and complexity ever since.

These are empirical facts, observable and measurable. Physics describes them brilliantly.

Philosophy is the step backward to see the whole design. It asks what makes the structure possible. Why is the universe relational rather than a collection of independent things? Why is it intelligible rather than chaotic? Why does it follow rational laws? Why did it begin in a state capable of producing galaxies, stars, planets, and eventually conscious beings who can understand it?

These questions point beyond physics to metaphysics, the study of being itself. And classical metaphysics has answers. The universe is contingent. Nothing in it contains the reason for its own existence. Every star, every law, every quantum field could have been otherwise. Contingent existence requires a necessary ground. It requires something whose essence is existence itself, something that cannot not be, something upon which all contingent things depend but which depends on nothing.[2]

Theology knows the Artist. It identifies this necessary ground as the God revealed in Scripture. Not as a being among beings, however powerful, but as Being Itself. The One who says "I AM." The eternal Logos through whom all things were made. The Trinity—Father, Son, and Spirit—who is fundamentally relational within His own divine life and creates a relational universe as the overflow of His love.

This God is not distant or abstract. He speaks, and creation comes into being. He establishes distinctions: light from darkness, sea from land, day from night. He names things and gives them their identity. He sustains the universe at every moment, not as a watchmaker who winds a clock and walks away, but as the continuous source of existence itself. He enters His own creation in Jesus Christ, the Word made flesh, revealing that the rational structure of the cosmos and the personal love of God are not two different things but one.[3]

The (IAM) shows how these three perspectives—physics, philosophy, theology—converge without collapsing into one another.

Consider a few key features of the universe and see how all three perspectives illuminate them:

Physics observes: Quantum entanglement binds particles in immediate correlation across vast distances. Particles are defined by their interactions, not in isolation. Space itself emerges from relational patterns.

Philosophy asks: Why is reality fundamentally relational rather than atomistic? What does this say about the nature of being?

Theology reveals: God is Trinity, three persons in eternal communion. Relationality is not accidental but reflects the nature of the Creator. "To be" means "to be in relation."[4]

The universe is intelligible.

Physics observes: Mathematical laws govern physical processes with stunning precision. The same equations work everywhere, from quarks to quasars.

Philosophy asks: Why is the universe comprehensible to minds? Why does mathematics work? What correspondence exists between thought and reality?

Defining the Informational Actualization Model

Theology reveals: The universe is created by the Logos, divine Reason itself. Intelligibility is built in because the cosmos is the work of a rational God who creates rational beings in His image.[5]

The universe generates increasing complexity.

Physics observes: Entropy increases, but local order emerges. Stars form, galaxies organize, chemistry becomes complex, life appears, consciousness awakens.

Philosophy asks: Why does the universe move from simplicity toward complexity? What directs this trajectory?

Theology reveals: Creation unfolds as God intends. The "formless void" becomes a cosmos. Potentiality becomes actuality. The universe is not static but dynamic, invited to participate in its own becoming.

Conscious beings emerge who can understand the universe.

Physics observes: Matter organizes into brains capable of processing information, integrating experience, and reflecting on existence.

Philosophy asks: Why does the universe produce observers? What does consciousness reveal about reality?

Theology reveals: Human beings are made in the image of God: rational, relational, free, capable of knowing truth and loving goodness. Consciousness is not an accident but the universe awakening to its Creator.[6]

Do you see the pattern? Physics describes. Philosophy explains what makes description possible. Theology identifies the source.

None of these levels eliminates the others. You don't stop needing physics when you have philosophy. You don't stop needing philosophy when you have theology. Each is essential. Each respects the boundaries of the others. And together, they reveal a universe far richer than any single perspective could capture.

Before going further, let's be absolutely clear about what the (IAM) does not claim. This matters, because misunderstanding these boundaries has caused needless conflict.

This is not an argument that physics proves God exists. Physics describes how the universe works. It cannot prove or disprove metaphysical claims. The (IAM) shows coherence, not proof. It demonstrates that the universe revealed by physics is compatible with the God revealed in Scripture, but compatibility is not the same as demonstration.

This is not a claim that the Bible is a science textbook. Scripture tells us **who** created, **why** He created, and **what** creation means. It does not tell us the age of the universe in years, the mechanics of stellar nucleosynthesis, or the equations of general relativity. Genesis is theology, not physics. Treating it as a competing scientific account misses its purpose entirely.

This is not an assertion that information is conscious or that the universe is a computer. Information in physics is mathematical and structural. It describes distinctions, correlations, and constraints. It is not semantic. It does not think. It does not mean anything by itself. When we say the universe can be described using information theory, we mean the mathematical framework is useful, not that reality is literally made of data.[7]

This is not an attempt to collapse domains. Physics remains physics. Philosophy remains philosophy. Theology remains theology. Each has its own methods, its own questions, and its own answers. The synthesis respects these boundaries. It shows how they relate without confusing them.

The (IAM) invites you to see the universe as you might see elegant code: structured, intelligible, purposeful. It does not ask you to reject science. It asks you to recognize that science describes the structure, philosophy asks what grounds it, and theology identifies the rational, relational Creator who brought it into being.

Think of it this way. Imagine three people examining the same landscape. One uses a topographical map and traces elevation changes, water flow, and geological features. Another considers why the landscape exists at all, what forces shaped it, and what purpose it serves. The third sees the landscape as the work of a Creator who intended it for habitation and flourishing.

All three perspectives are true. None contradicts the others.

Together, they provide a richer understanding than any single view could offer.

That is what the (IAM) proposes for reality itself. Physics maps the structure. Philosophy asks what grounds it. Theology reveals the personal source. The convergence is not proof, but it is coherence. And that matters because coherence, as this book will show, is not a weakness of explanation, but its deepest strength.

If you're a scientist, this framework respects your discipline. It does not ask you to bend empirical facts to fit theology. If you're a philosopher, this framework honors careful reasoning about being and existence. It does not collapse metaphysics into physics. If you're a theologian, this framework takes Scripture seriously. It does not reduce God to an abstraction or creation to mere mechanism.

The (IAM) is an invitation to see how these three ways of knowing illuminate one reality. It's a recognition that truth is unified, even when our disciplines are distinct. It's an acknowledgment that the smartest people in history like Newton, Kepler, Lemaître, and countless others all held all three perspectives together without contradiction. You don't have to choose between science and faith. You never did.

The remaining chapters explore this convergence in detail. You'll see how creation unfolds as the self-expression of the eternal I AM, why imperfection and freedom are necessary for genuine love, how the Logos connects divine speech with cosmic order, and why the relational God of Christian theology provides the ultimate ground for human dignity, meaning, and purpose.

This is not an argument that compels belief. It is an invitation to see how physics, philosophy, and theology, if each understood properly, illuminate a single coherent reality. The universe is rational because it comes from the Logos. It is relational because it flows from the Trinity. It is intelligible because it reflects the mind of God. And you, as a conscious being capable of grasping these truths, are part of the story the universe is telling.

The next chapter explores what it means for creation to unfold as the self-expression of the One who is I AM the God whose very name is Being Itself.

CHAPTER REVIEW:

KEY CONCEPTS:

- **The (IAM) proposes coherence, not proof.** Showing that physics, philosophy, and theology illuminate one reality when each operates properly

- **(IAM)** = Informational Actualization Model Echoing God's self-revelation in Exodus 3:14 while describing reality's unfolding.

- **The universe grows in informational content** - Moving from "formless void" toward ever-greater structure and meaningful distinction.

- **Three perspectives, one reality** - Physics (structure), philosophy (ground), theology (personal source).

- **What this is NOT** - Not physics proving God; not Bible as science textbook; not information as conscious; not theology from equations

WHAT THIS MEANS:

Informational = (does not mean computational or digital in a literal sense) It names the ordered, relational intelligibility of reality—the fact that the universe is structured, knowable, and non-chaotic in a way that invites understanding.

Actualization = names the metaphysical depth of the structure: the movement from possibility to reality, from potency to act, from what could be to what is. Reality is not self-originating. It is continually given, sustained, and held.

Model = means that it is not a closed system or final explanation, but a framework that allows physics, philosophy, and theology to illuminate one reality without collapsing into one another.

When physics, metaphysics, and theology each operate within their proper domains, they reveal complementary truths about a single coherent reality. The convergence is not coincidental but reflects the deep unity of truth itself. This framework respects disciplinary boundaries while showing how they illuminate each other.

IMPORTANT TERMS & CONCEPTS:

- **I AM** (divine name) → See Chapter 16; Part X, Section 1.1 (Theology domain)
- **Actualization** → Part X, Section 1.6 (Act vs. Potency; Actualization)
- **Coherence vs. Proof** → Part X, Section 3 (What This Work Does Not Claim)
- **Domain Boundaries** → Part X, Section 1 (all three domains explained)
- **Synthesis** → Part X, Section 7 (Integration Guidelines)

LOOKING AHEAD:
- Chapter 11 shows how creation unfolds as divine self-expression
- Chapter 12 addresses why imperfection and freedom are necessary
- Part IV connects this to Scripture (Genesis, John 1, Logos theology)
- Part V develops the relational nature of ultimate reality

TECHNICAL REFERENCE:
- For comprehensive statement of what is NOT claimed: Part X, Section 3 (entire section)
- For domain separation principles: Part X, Sections 1 & 2
- For how analogies work: Part X, Section 7.5 (Analogical Interpretation)
- For comparison with similar frameworks: Part X, Section 8 (Comparative Frameworks)

---

1. The divine name revealed to Moses in Exodus 3:14 has been understood throughout Jewish and Christian tradition as a statement about God's being. The Hebrew *'ehyeh 'asher 'ehyeh* is traditionally translated "I AM WHO I AM" or "I WILL BE WHAT I WILL BE," emphasizing God's self-existence and eternal presence. For classical theological treatment: Aquinas, T. *Summa Theologica*, Part I, Question 13, Article 11. For contemporary philosophical analysis: Davies, B. (2004). *An Introduction to the Philosophy of Religion*. Oxford University Press, Chapter 2. The name reveals that God is not one being among many but the source of being itself.
2. The argument from contingency to a necessary being is one of the most enduring in classical metaphysics. For Aquinas's formulation (the "Third

Way"): Aquinas, T. *Summa Theologica*, Part I, Question 2, Article 3. For contemporary rigorous defenses: Pruss, A. R. (2006). *The Principle of Sufficient Reason: A Reassessment*. Cambridge University Press. Feser, E. (2017). *Five Proofs of the Existence of God*. Ignatius Press, Chapter 1. Rasmussen, J. (2019). *How Reason Can Lead to God*. InterVarsity Press. These arguments demonstrate that contingent existence points necessarily toward a ground of being whose essence is existence.

3. For the doctrine of the Logos in John's Gospel and its connection to both Jewish wisdom literature and Greek philosophy, see: Brown, R. E. (1966). *The Gospel According to John I–XII*. Anchor Bible Series. Doubleday, pp. 519–524. For the theological development: Torrance, T. F. (1988). *The Trinitarian Faith: The Evangelical Theology of the Ancient Catholic Church*. T&T Clark. The Johannine prologue (John 1:1–18) identifies Jesus Christ as the eternal Word through whom all things were made, uniting creation and incarnation.

4. The relational nature of God in Trinitarian theology and its implications for understanding reality has been explored extensively in modern theology. For foundational work: Zizioulas, J. D. (1985). *Being as Communion: Studies in Personhood and the Church*. St Vladimir's Seminary Press. For the connection to quantum relationality: Polkinghorne, J. (2005). *Exploring Reality: The Intertwining of Science and Religion*. Yale University Press, Chapter 4. The Trinity reveals that being itself is relational, not atomistic.

5. For the "unreasonable effectiveness of mathematics" and its philosophical implications, see: Wigner, E. (1960). "The Unreasonable Effectiveness of Mathematics in the Natural Sciences." *Communications in Pure and Applied Mathematics* 13(1), 1–14. For theological interpretation: Polkinghorne, J. (1998). *Belief in God in an Age of Science*. Yale University Press, Chapter 1. The correspondence between mathematical structures and physical reality suggests a rational ground for both.

6. The doctrine of the imago Dei (image of God) establishes that human beings uniquely bear God's image through rationality, relationality, moral agency, and creative capacity. For biblical foundation: Genesis 1:26–27. For theological development: Aquinas, T. *Summa Theologica*, Part I, Question 93. For contemporary treatment: Cortez, M. (2010). *Theological Anthropology: A Guide for the Perplexed*. T&T Clark, Chapter 2. Human consciousness reflects the rational and relational nature of the Creator.

7. For the critical distinction between syntactic information (physical) and semantic information (meaning), see Part X, Section 4 of this work. For technical treatment: Floridi, L. (2011). *The Philosophy of Information*. Oxford University Press, Chapter 4. Information theory in physics describes structural distinctions and correlations but does not carry intrinsic meaning or consciousness. This distinction prevents the error of treating the universe as literally computational or conscious.

CHAPTER ELEVEN
CREATION AS THE SELF-UNFOLDING OF THE ETERNAL I AM

To say that God is I AM is to say that God is Being Itself. Not a being among beings, however supreme. Not the first cause in a series of causes, however powerful. But the sheer act of existence from which all particular existences flow. God does not have being. God is being. Everything else that exists participates in being but does not possess it essentially. This is the classical doctrine of divine simplicity, taught by theologians from Augustine to Aquinas to the present. God's essence and God's existence are identical. In God alone, to be and to exist are one and the same.[1]

This understanding transforms how you think about creation. If God is Being Itself, then creation is not the manufacture of objects from raw materials. It is not divine craftsmanship working with pre-existing stuff. It is the granting of existence to what otherwise would not be. Creation is the act by which God shares being with what is not God. It is the overflow of divine generosity, the gift of participation in the one act of existence that is God Himself.

This is what theologians mean by **creatio ex nihilo**, creation from nothing. The phrase is often misunderstood. It does not mean that "nothing" is a kind of raw material from which God fashioned the world, as if nothing were a shadowy substance waiting to be shaped. It means that before creation, there was only God. There was no space, no time, no matter, no energy, no potential waiting to be actualized. There was only the infinite fullness of divine being. Creation is the act by which God brings into existence what is entirely other than Himself, not by transforming something that already existed, but by freely willing that it should be.[2]

Scripture expresses this through the imagery of divine speech. God said, "Let there be light," and there was light (Genesis 1:3, NABRE). God said, "Let the earth bring forth vegetation," and it did. God said, "Let us make man in our image," and humanity came into being. The pattern is consistent: God speaks, and reality responds. The Word goes forth, and creation unfolds. This is not mere poetic language. It is a theological claim about the nature of creation. The universe exists because God speaks it into existence. And God's speech is not like human speech, a vibration of air carrying information from one mind to another. God's speech is creative. It brings into being what it names. The Word of God is efficacious. It accomplishes what it declares.[3]

This connection between divine speech and creation reveals something profound about the nature of reality. If the universe comes into being through the Word, then the universe is inherently meaningful. It is not brute matter obeying blind forces. It is the expression of divine intention. Every structure, every law, every pattern reflects the rationality of the Logos through whom all things were made. The intelligibility of the cosmos is not an accident. It is a consequence of the fact that the cosmos is the product of divine Reason.

Modern physics, without intending to, describes a universe that fits this pattern remarkably well. The early universe began in a state that was remarkably simple yet profoundly ordered. From that initial state, structure emerged. Particles formed. Forces differentiated. Matter clumped under gravity. Stars ignited. Galaxies assembled. The progression was not random. It followed laws. Those laws permitted the development of complexity. And that complexity eventually gave rise to conscious beings capable of understanding the laws themselves. The universe, in a sense, learned to know itself.

Theology describes this same progression in different language. Creation unfolds as God brings order from formlessness, distinction from undifferentiation, actuality from potentiality. Genesis portrays this unfolding as a sequence of separations: light from darkness, waters above from waters below, sea from land, day from night. Each distinction is meaningful. Each separation generates a new dimension of order. The formless void becomes a structured cosmos capable of sustaining life. This is not mechanism. It is the language of purpose and design. God intends creation to be ordered, and it is. God intends creation to be fruitful, and it becomes so.

The concept of actualization captures this dynamic. In classical metaphysics, every finite thing is a composite of actuality and potentiality. Actuality is what a thing is at present. Potentiality is what it could become. A seed is actually a seed, but it is potentially a tree. An electron is in one location, but it could be in another. The universe itself, at any given moment, is actually in one state, but it contains potentialities for countless future states. Actualization is the process by which potentials become actual, by which possibilities are realized, by which what could be becomes what is.[4]

God is pure actuality. In God, there is no potentiality, no unrealized possibility, no capacity to become something other than what He already is. God is complete, perfect, and unchanging. Creation, by contrast, is a mixture of actuality and potentiality. And the unfolding of creation is the progressive actualization of potentials under the sustaining providence of God. Stars form because the potential for stellar fusion is actualized in collapsing gas clouds. Life emerges because the potential for self-replication and complexity is actualized in chemical systems. Consciousness arises because the potential for awareness is actualized in sufficiently complex organisms. At every stage, what was potential becomes actual. And this actualization is not arbitrary. It follows patterns, obeys laws, and moves toward greater complexity and order.

This is not to say that God micromanages every event or that natural processes are illusions. The doctrine of secondary causation affirms that created things genuinely act. Gravity really pulls. Chemical reactions really occur. Organisms really reproduce. These are not divine puppetry. They are real causal powers that God has given to creation. But secondary causes depend on the primary cause. Gravity operates because God sustains the structure of spacetime. Chemicals react because God maintains the laws of quantum mechanics. Organisms reproduce because God upholds the stability of biological information. Creation has genuine causal efficacy, but that efficacy is participated, not autonomous. It is a gift from the One who is the source of all causation.[5]

The progression from simplicity to complexity, from potentiality to actuality, from formlessness to order is not an accident. It reflects the character of the Creator. God is not chaotic. God is supremely ordered. The Trinity is perfect harmony, three persons in one essence, eternal communion without confusion or division. This divine order is reflected in creation. The laws of physics are not arbitrary. They express the rationality of the Logos. The fine-tuning of constants is not luck. It reflects divine intention. The emergence of conscious beings capable of relationship, choice, and moral responsibility is not a cosmic fluke. It is the fulfillment of a purpose woven into creation from the beginning.

God creates through the Word, and the Word establishes order. This is why the universe is intelligible rather than chaotic, structured rather than formless, capable of sustaining life rather than hostile to it. The very existence of natural law points toward a Lawgiver. The very fact that patterns repeat, that mathematics describes nature, that experiments yield consistent results: all of this presupposes a fundamental rationality built into the fabric of reality. And that rationality traces back to the Logos, the divine Reason, the second person of the Trinity through whom all things were made.

Creation as the Self-Unfolding of the Eternal I AM   123

Creation is also relational from the start. God is Trinity—Father, Son, and Holy Spirit—three persons in eternal communion. Relationality is not something God possesses. It is what God is. And this relational nature is reflected in creation. Quantum entanglement binds particles in immediate correlation. Space emerges from patterns of relationship. Information is defined by distinctions that exist in relation to one another. Even at the most fundamental level, reality is structured relationally, not atomistically. To be is to be in relation. This is not pantheism. It is the recognition that the relational God creates a relational universe.[6]

Human beings occupy a unique place in this unfolding. You are made in the image of God. This does not mean physical resemblance but rational capacity, moral agency, relational nature, and the ability to know and love. You are called to exercise stewardship over creation, to cultivate it, to understand it, and to name it. Naming is not arbitrary labeling. It is the recognition of what things are, the acknowledgment of their nature and purpose. In Genesis, Adam names the animals. In modern science, we name species, laws, forces, and particles. Both are acts of understanding, acts of participating in the rationality of the Logos.

God creates beings capable of genuine choice not because He lacks control but because love requires freedom. A creature that is forced to obey, that has no capacity for refusal, is not capable of love. Love is the willing of the good for another, and willing presupposes choice. God desires creatures who can freely enter into relationship with Him, who can choose to love and be loved, who can participate consciously and willingly in the divine life. This means granting freedom. And freedom necessarily includes the possibility of misuse. The risk of evil is the price of love.[7]

But God does not abandon creation to the consequences of freedom misused. The doctrine of redemption affirms that God enters creation to restore what has been damaged, to heal what has been broken, to reconcile what has been estranged. The Incarnation is the ultimate expression of this. The Logos through whom all things were made becomes incarnate in Jesus Christ. The Word that spoke the universe into being takes on human flesh, enters history, suffers, dies, and rises again. This is not a divine intervention from outside the system. It is the Creator entering His own creation to bring it to fulfillment. Redemption is the completion of creation. It is the final actualization of the potential for which the universe was made: union with God.

The unfolding of creation, then, is not a mechanical process. It is the story of God's self-giving love. It is the narrative of being shared, of potentials actualized, of order emerging from chaos, of life arising from matter, of consciousness awakening in flesh, of persons coming into relationship with the One who made them. Every star that forms, every organism that reproduces, every person who asks why: all of this is part of the single, continuous act by which God gives existence to what is not God and invites it to share in the divine life.

This vision transforms how you understand your place in the cosmos. You are not an accident. You are not a meaningless arrangement of particles obeying impersonal laws. You are a participant in the self-expression of the eternal I AM. You are a creature brought into being by the Word, sustained by the Word, and called to know the Word. Your capacity to think, to choose, to love, to create—all of this reflects the image of God within you. And your longing for meaning, for truth, for goodness, for beauty is the echo of the divine calling you back to the Source from which you came.

The next chapter explores why this calling includes risk, why love requires the possibility of refusal, and why the imperfections you observe in the world do not negate the goodness of the Creator but reveal the depth of His commitment to freedom and relationship.

CHAPTER REVIEW:
KEY CONCEPTS:

- **Creation ex nihilo** - God creates from nothing, not from preexisting material; before creation, only God existed

- **Divine speech is efficacious** - God's Word accomplishes what it declares; creation is the expression of divine intention

- **Creation unfolds through meaningful distinctions** - Genesis portrays separation (light/darkness, sea/land) as generating order
- **Actualization of potential** - The universe moves from simple initial state toward complexity through divine purpose
- **Relational structure throughout** - From quantum entanglement to human persons, reality is fundamentally relational

WHAT THIS MEANS:

Creation is not mechanism but the overflow of divine generosity. The universe exists because God freely wills it into being through the Word (Logos). The informational/relational structure revealed by physics aligns with theological claims about creation through divine speech establishing meaningful distinctions. This is coherence, not derivation.

IMPORTANT TERMS & CONCEPTS:

- **Creatio ex nihilo** → Part X, Section 1.1 (Theology); Q&A Section II
- **Davar** (Hebrew: word/event) → See Chapter 13; Part X theological terms
   Divine Speech → Part X, Section 1.1 (Theology domain)
- **Potency to Act** → Part X, Section 1.6 (Actualization)
- **Logos** → See Chapter 14; Part X, Section 1.1 (Theology)

LOOKING AHEAD:

- Chapter 12 explains why genuine freedom requires risk and imperfection
- Chapter 13 examines Genesis as information sequence
- Chapter 14 develops Logos theology connecting reason and creation
- Part V explores how the Triune God is the ground of relational reality

TECHNICAL REFERENCE:

- For creatio ex nihilo properly understood: Part X, Q&A Section II, Questions 11-13
- For what "Word" means theologically vs. linguistically: Part X, Section 7.5 (Analogical Interpretation)

- For primary vs. secondary causation: Part X, Section 1.6 (Causality concepts)
- For Genesis interpretation: See Chapter 13 endnotes

---

1. The doctrine of divine simplicity holds that God is not composed of parts, properties, or attributes that could be distinguished from His essence. For classical formulation: Aquinas, T. *Summa Theologica*, Part I, Question 3. For contemporary philosophical defense: Dolezal, J. E. (2011). *God Without Parts: Divine Simplicity and the Metaphysics of God's Absoluteness*. Pickwick Publications. Davies, B. (2004). *An Introduction to the Philosophy of Religion*. Oxford University Press, Chapter 2. Divine simplicity prevents us from thinking of God as a complex being whose parts could be separated or analyzed.
2. For the doctrine of **creatio ex nihilo** and its biblical and theological foundations, see: Copan, P., & Craig, W. L. (2004). *Creation Out of Nothing: A Biblical, Philosophical, and Scientific Exploration*. Baker Academic. May, G. (1994). *Creatio Ex Nihilo: The Doctrine of 'Creation out of Nothing' in Early Christian Thought*. T&T Clark. The doctrine affirms that God created freely, not from necessity or from pre-existing materials, establishing the absolute dependence of all creation on God.
3. For the theology of divine speech and its creative power in Scripture, see: Wolterstorff, N. (1995). *Divine Discourse: Philosophical Reflections on the Claim that God Speaks*. Cambridge University Press. For the Hebrew concept of **davar** (word as event), see: Boman, T. (1960). *Hebrew Thought Compared with Greek*. SCM Press, pp. 58–69. God's word in Scripture is performative—it accomplishes what it declares (Isaiah 55:11).
4. The distinction between actuality and potentiality is fundamental to Aristotelian-Thomistic metaphysics. For Aristotle's original treatment: Aristotle. *Metaphysics*, Book IX (Theta). For Aquinas's development: Aquinas, T. *Summa Theologica*, Part I, Questions 3–4. For contemporary exposition: Feser, E. (2014). *Scholastic Metaphysics: A Contemporary Introduction*. Editiones Scholasticae, Chapters 1–2. This framework explains change, becoming, and the actualization of possibilities in creation.
5. The doctrine of primary and secondary causation distinguishes between God's immediate action (primary cause) and the genuine causal powers of creatures (secondary causes). For classical treatment: Aquinas, T. *Summa Contra Gentiles*, Book III, Chapters 66–70. For contemporary discussion: Dodds, M. J. (2012). *Unlocking Divine Action: Contemporary Science and Thomas Aquinas*. Catholic University of America Press. This doctrine preserves both divine sovereignty and creaturely agency without contradiction.
6. For the relational nature of the Trinity and its implications for creation, see: Zizioulas, J. D. (1985). *Being as Communion: Studies in Personhood and the Churcc*. St Vladimir's Seminary Press. For the connection between Trinitarian theology and quantum relationality: Polkinghorne, J. (2005). *Exploring Reality: The Intertwining of Science and Religion*. Yale University Press, Chapter 4. LaCugna, C. M.

(1991). *God For Us: The Trinity and Christian Life*. HarperSanFrancisco. The Trinity reveals that being itself is fundamentally relational, not atomistic.

7. For the classical argument that love requires freedom and the theological problem of evil, see: Plantinga, A. (1974). *The Nature of Necessity*. Oxford University Press, Chapter 9. Swinburne, R. (1998). *Providence and the Problem of Evil*. Oxford University Press. For the patristic perspective: Irenaeus of Lyons. *Against Heresies*, Book IV, Chapters 37–38 (the soul-making theodicy). These works explore why a good God would create beings capable of genuine choice, even at the risk of evil.

CHAPTER TWELVE
IMPERFECTION, FREEDOM, AND THE PURPOSE OF RISK

*I*f creation is the unfolding of divine generosity, grounded in perfect wisdom and sustained by the eternal act of being, then a profound question emerges: why is the world imperfect? Why does it contain suffering, limitation, uncertainty, and moral failure? Why would a perfect God create a universe in which creatures can refuse the good, misunderstand the truth, and wound themselves and others?

These questions are not new. They stand at the center of theological and philosophical reflection across cultures and centuries. Yet they gain new depth when considered in light of a relational universe. The structure of creation itself suggests that imperfection and risk are not intrusions into the divine plan but necessary conditions for the existence of freedom, agency, and authentic personhood.

To begin, you must distinguish between two kinds of imperfection. Metaphysical imperfection refers to finitude, the simple fact that created beings are limited. A stone cannot think. A plant cannot make moral choices. An animal cannot contemplate its own existence. None of this constitutes evil. It is simply the nature of creation. Only God possesses infinite fullness. Everything else participates in being to the degree that its nature allows. This kind of limitation is not a flaw but a necessary feature of anything that is not God. To be created means to be finite. To be finite means to have boundaries, to exist in one way rather than all ways.[1]

Moral imperfection, by contrast, arises when a rational creature chooses contrary to the good. This kind of imperfection is not created by God. It is made possible by freedom. Any universe containing creatures capable of rational agency necessarily contains the possibility of error, misuse, and disorder. Freedom is the condition for moral responsibility, and responsibility exists only where real choice exists. You cannot praise someone for doing the good if they could not have done otherwise. You cannot hold someone accountable for wrongdoing if they had no genuine alternative.

Why, then, would God create a world where such risks exist? The answer begins with the nature of love. Love cannot be compelled. It cannot be engineered. It cannot be guaranteed. Love is possible only where freedom exists. Scripture declares not merely that God loves, but that God is love. This doesn't mean that God possesses love as one quality among others, but that love is essential to who He is. The Trinity is eternal love: Father, Son, and Holy Spirit giving and receiving in perfect communion.[2] Love, in its essence, is the willing of the good for another. It is relational, self-giving, and creative. It does not seek its own but delights in the flourishing of the beloved.

Love requires freedom because forced affection is not love at all. A relationship without the possibility of refusal is not a relationship but compulsion, control, or programming. Therefore, if God desires communion with His creatures. If He calls them not to be servants but friends, not puppets but persons. He must endow them with free will. Freedom is not a flaw or an unfortunate necessity. It is a gift necessary for the fulfillment of love. Without freedom, there can be no genuine relationship. Without genuine relationship, there can be no love.[3]

The universe reflects this truth in its structure. As established in

Chapter 5, physics describes mechanisms, metaphysics explains being, and theology addresses ultimate cause. Within these proper domains, we can observe how the relational nature of creation allows for genuine choice. The physical universe is not a closed, deterministic machine grinding through predetermined states. Quantum events exhibit genuine indeterminacy. Biological processes allow for variation and adaptation. Human decisions involve real deliberation and choice. This openness is not chaos. It is the space in which love can flourish, in which persons can grow, and in which freedom can be exercised toward the good.[4]

The freedom of the human will extends from the relational structure of reality itself. You are not isolated atoms operating in mechanical independence. You are persons embedded in relationships, capable of genuine agency within a world that responds to your choices. Every decision you make ripples through the web of relationships that constitutes human life. You affect others. You shape communities. You leave marks on history. This is what it means to be free in a relational universe.

But freedom brings risk. The possibility of choosing the good necessarily implies the possibility of choosing evil. This is not a defect in creation. It is the logical consequence of granting freedom. A being capable of love must also be capable of refusing love. A being capable of truth must also be capable of embracing falsehood. A being capable of justice must also be capable of committing injustice. To remove these possibilities would be to remove freedom itself, and with it the possibility of genuine relationship.

Scripture presents this tension from the beginning. Humanity is placed in a world where obedience and disobedience are both possible. Love is offered, but it is not coerced. The moral structure of the universe reflects this reality. Actions have consequences. Virtue leads to harmony, flourishing, and peace. Vice leads to disorder, suffering, and alienation. The pattern is woven into creation itself, not as arbitrary punishment but as the natural outworking of choices in a relational universe. When you act in accordance with the good, you align yourself with the structure of reality as it is meant to be. When you act contrary to the good, you introduce dissonance, disruption, and damage.

This understanding does not diminish the seriousness of evil. It does not excuse suffering or make light of human wrongdoing. Rather, it places these realities in a framework where their existence is neither a sign of divine weakness nor evidence against a rational Creator. It shows that the possibility of imperfection follows necessarily from the gift of freedom, and that the gift of freedom is necessary for the existence of personal beings capable of communion with God.

From a theological perspective, God creates not because He lacks anything but because His nature is generosity. Generosity always involves giving space. To give freely is to allow the other to be what the other is. To create a free being is to accept that this being may choose the good imperfectly. True freedom includes the ability to misuse freedom. The very capacity that allows the human soul to love God also allows it to turn away. The capacity that allows for greatness also allows for failure. This is not a defect in creation but the logic of freedom.[5]

There is another dimension to this question. Why did God create a world in which finite creatures must learn? Why not create beings fully formed, complete, and incapable of error? The answer lies again in the nature of relationship and personhood. A being that simply possesses knowledge without having learned it has not developed. A being that cannot grow cannot mature. Growth is essential to personhood. Creation is designed as a journey. Human beings come into the world incomplete and are invited to participate in their own becoming.

The universe itself mirrors this trajectory. It begins in simplicity and moves toward complexity. It begins in silence and unfolds into intelligibility. It begins as potential and blossoms into persons. This arc reflects divine intention. Creation is not static because God intends for creatures to walk a path toward fullness. Time allows for development. History allows for narrative. Freedom allows for meaning. The universe is not merely a place where beings exist. It is a place where beings grow, where they learn, where they become what they were created to be through the exercise of genuine choice in a genuinely open world.[6]

Imperfection also serves another purpose. It teaches dependence. A world of ease would obscure the fact that all being is gift. The challenges of existence lead creatures to seek wisdom beyond themselves. Difficulty invites humility. Suffering awakens compassion. Limitations drive longing for the One who is not limited. Moral imperfection reveals the need for grace. In this sense, imperfection does not oppose divine intention but becomes a channel through which creatures discover the depth of their need for God. The journey from potential to actuality, from ignorance to understanding, from self-centeredness to love. This is the story of human life, and it requires struggle, choice, and the ever-present possibility of failure.

The (IAM) gives a unique perspective on these questions. If creation is the unfolding of divine generosity, then freedom is woven into the fabric of reality itself. Quantum openness, biological variability, rational agency, and moral responsibility form a single continuum of increasing freedom. The universe is designed not as a perfectly predictable machine but as a relational universe in which genuine novelty, choice, and love are possible. The informational structure described in earlier chapters (see Part X, Section 8 for detailed treatment) reveals a cosmos that is both ordered and open, both lawful and free, both deterministic in its general patterns and indeterminate in its particular actualizations.

This does not make God the author of moral evil. Rather, it shows that moral evil arises from the misuse of a good gift. God provides the ground of being. Creatures provide the use of their freedom. God sustains the world in existence at every moment through the continuous act by which He shares being with what is not God. Creatures determine how they act within that sustained existence. God establishes the structure of reality in which moral choice becomes meaningful. Creatures make the choices. The possibility of failure is not the purpose of creation, but its unavoidable condition given the purpose that creatures be genuinely free.

Finally, the Christian tradition teaches that God does not abandon creatures in their imperfection. The Logos enters creation not merely as its foundation but as its redemption. The God who gives being also gives restoration. The One who sustains the universe also enters its history. The Word through whom all things were made becomes incarnate in Jesus Christ, taking on human flesh, entering time, suffering under the weight of human brokenness, dying, and rising again to complete what creation began. This is not a divine intervention from outside the system, as if God were a mechanic fixing a broken machine. It is the Creator entering His own creation to bring it to fulfillment. Redemption is the completion of creation. It is the final actualization of the potential for which the universe was made: union with God.[7]

The risk of freedom, therefore, is met by the greater depth of divine love. God does not shield His creatures from every consequence of their choices. To do so would be to remove the very freedom that makes them persons. Instead, He enters the suffering that freedom makes possible and transforms it from within. He takes the worst that human freedom can produce—betrayal, injustice, violence, death—and makes it the means of redemption. The cross is not Plan B. It is the revelation of how deeply God commits to freedom and how far He will go to restore what has been broken without violating the dignity of the creatures He has made.

In summary, the universe contains imperfection because it contains persons. It contains risk because it contains freedom. It contains struggle because it contains growth. These realities do not compete with the goodness of creation. They arise from the very generosity that makes creation meaningful. A world without freedom could not reveal the fullness of divine love nor allow creatures to become what they were created to be. A world without risk would be a world without choice. A world without choice would be a world without love. And a world without love would be no reflection of the God who is love.

You are not an accident. You are not a meaningless arrangement of particles obeying impersonal laws. You are a participant in the self-expression of the eternal I AM. You are a creature brought into being by the Word, sustained by the Word, and called to know the Word. Your capacity to think, to choose, to love, to create—all of this reflects the image of God within you. And your longing for meaning, for truth, for goodness, for beauty is the echo of the divine calling you back to the Source from which you came. The imperfections you experience, the choices you face, the struggles you endure—all of these are part of the journey toward fullness, the path from potential to actuality, from creature to participant in the divine life.

The next chapter turns from these metaphysical questions to Scripture, beginning with the opening lines of Genesis. If the universe unfolds through relational structure, what does it mean that God creates through speech? What can the ancient words "Let there be" tell us about the deep nature of reality? How does the informational framework we have explored connect to the testimony of Scripture about the beginning of all things?

CHAPTER REVIEW:

KEY CONCEPTS:

- **Metaphysical imperfection (finitude) vs. moral imperfection (evil)** - Created things are limited by nature; this is not evil but the nature of being created

- **Love requires freedom** - Coerced or programmed "love" is not genuine; authentic relationship requires the possibility of refusal

- **Freedom entails genuine risk** - A world without the possibility of wrong choice would be a world without love

- **Evil as privation of good** - Not a positive force but absence/corruption of what should be present

- **Creation permits participation** - God doesn't create automatons but genuine agents who can cooperate or resist

WHAT THIS MEANS:

The existence of imperfection and the possibility of evil are not design flaws, but necessary features of a creation intended for genuine relationship. A perfect God creates free beings precisely because love, the deepest purpose of creation, cannot exist without freedom. The risk is real, but so is the reward: authentic communion between Creator and creature.

IMPORTANT TERMS & CONCEPTS:

- **Evil as Privation** → Part X, Section 1.6 (classical metaphysics); Q&A Section IV

- **Free Will** → Part X concepts; Q&A Section IV, Questions 31-40

- **Metaphysical vs. Moral Imperfection** → Part X, Section 1.6 (essence, finitude)

- **Divine Permission of Evil** → Q&A Section IV

- **Soul-Making Theodicy** → Referenced in Chapter 12 endnotes

LOOKING AHEAD:

- Part IV (Chapters 13-15) connects this to biblical narratives of creation and redemption

- Part V develops how the relational God is the ground of genuine love

- Q&A Section IV addresses specific questions about evil and suffering

- Chapter 22 explores human response to divine invitation despite brokenness

TECHNICAL REFERENCE:

- For comprehensive theodicy treatment: Part X, Q&A Section IV (Questions 31-40)

- For freedom and divine sovereignty: Part X, Section 5 (Anticipated Objections)

- For privation theory of evil: Part X, Section 1.6 (classical metaphysics concepts)

- For what this does NOT claim about evil: Part X, Section 3

---

1. The distinction between metaphysical and moral imperfection is fundamental to classical theodicy. For metaphysical finitude as a necessary feature of created being: Aquinas, T. *Summa Theologica*, Part I, Question 48, Articles 1–3. Davies, B. (2006). *The Reality of God and the Problem of Evil*. Continuum, Chapter 2. Evil as

privation (absence of good) rather than positive reality is a key Augustinian and Thomistic insight.

2. For the theological foundation that God's essence is love: 1 John 4:8, 16. Systematic treatments: Zizioulas, J. D. (1985). *Being as Communion: Studies in Personhood and the Church*. St Vladimir's Seminary Press. LaCugna, C. M. (1991). *God For Us: The Trinity and Christian Life*. HarperSanFrancisco. The doctrine that God is love (not merely possesses love) grounds the understanding of creation as self-giving generosity.

3. The classical free will defense in theodicy: Plantinga, A. (1974). *The Nature of Necessity*. Oxford University Press, Chapter 9 (on God, freedom, and evil). Swinburne, R. (1998). *Providence and the Problem of Evil*. Oxford University Press. These works argue that morally significant free will requires the genuine possibility of both good and evil choices. For critique and refinement: Adams, M. M., & Adams, R. M. (Eds.). (1990). *The Problem of Evil*. Oxford University Press.

4. For quantum indeterminacy and its philosophical implications: Heisenberg, W. (1958). *Physics and Philosophy: The Revolution in Modern Science*. Harper & Row, Chapter 3. For biological openness and variability: Mayr, E. (2001). *What Evolution Is*. Basic Books, Chapters 6–7. These provide the scientific basis for understanding creation as genuinely open rather than mechanistically determined. See also Part X, Section 8 for treatment of determinism, indeterminacy, and freedom.

5. For the logic of free will and divine creation: Lewis, C. S. (1940). *The Problem of Pain*. Macmillan, Chapter 2 (on divine omnipotence and logical impossibility). Stump, E. (2010). *Wandering in Darkness: Narrative and the Problem of Suffering*. Oxford University Press. Lewis argues that even omnipotence cannot create logically impossible states (e.g., free beings who cannot choose evil).

6. The soul-making theodicy, developed from Irenaeus: Hick, J. (1966). *Evil and the God of Love*. Harper & Row, Part IV. This tradition emphasizes that creation is designed for growth, development, and the formation of character through genuine struggle and choice. For critique and development: Meister, C., & Dew, J. K. (Eds.). (2013). *God and Evil: The Case for God in a World Filled with Pain*. IVP Academic.

7. For the Incarnation as completion of creation: Torrance, T. F. (1981). *Divine and Contingent Order*. Oxford University Press. Wright, N. T. (2003). *The Resurrection of the Son of God*. Fortress Press, Chapter 18 (on redemption completing creation). Athanasius, *On the Incarnation* (4th century). The patristic and Reformed traditions both emphasize that redemption is not a divine afterthought but the intended fulfillment of creation from the beginning. For the cross as revelation of divine love within suffering: Moltmann, J. (1974). *The Crucified God*. SCM Press.

PART FOUR
THE LOGOS: WHERE PHYSICS MEETS SCRIPTURE

CHAPTER THIRTEEN
GENESIS AS AN INFORMATION SEQUENCE

When the opening lines of Genesis declare that God creates through speech, they offer one of the most profound statements about reality ever written. The words "And God said, 'Let there be light,' and there was light" (Genesis 1:3, ESV) have shaped centuries of theological reflection. Yet their depth becomes even more striking when read alongside what modern physics reveals about the relational structure of the universe. Genesis portrays creation not as a battle, accident, or mechanical process, but as an ordered sequence of distinctions spoken into existence by a rational, personal God. It presents creation as the articulation of meaning.

This chapter examines the creation account of Genesis 1 as an informational sequence, a divine ordering of potential into actual form through meaningful acts of speech. The purpose is not to turn Scripture into physics or physics into theology, but to show that the conceptual structure of Genesis aligns in surprising ways with the informational and relational universe described by modern science. As established in Chapter 5, these domains remain distinct: physics describes mechanisms, metaphysics explains being, theology addresses ultimate cause. Yet when each operates within its proper domain, they reveal a coherent vision of reality.

The first thing you notice is that the biblical creation narrative begins not with matter but with formless potential. "The earth was without form and void" (Genesis 1:2, ESV)—a phrase that conveys unordered possibility rather than absolute nonexistence. In ancient Hebrew, the expression **tohu wabohu** describes a state lacking distinction, pattern, or ordered separation. Something exists, but it does not yet have identity or structure. The phrase speaks of potentiality waiting for actualization, of undifferentiated existence waiting to be given form.[1] This mirrors a concept found throughout physics: systems can exist in states of superposed possibility before distinctions are established. A quantum system before measurement contains multiple potential outcomes. The early universe, though intensely hot and dense, possessed a kind of simplicity—a state from which all the complexity we observe would eventually emerge. The Genesis account captures this intuition: creation begins with potential that God then actualizes through a sequence of meaningful distinctions.

This is not to say that Genesis describes quantum mechanics or that quantum mechanics validates Genesis. Rather, both accounts recognize that ordered reality emerges from something less ordered, that form arises from formlessness, that structure develops from potential. One describes this theologically as the creative work of God. The other describes it physically as the evolution of the cosmos under natural law. But the conceptual pattern resonates: distinction precedes order, and order precedes complexity.

The second feature of Genesis is its repeated emphasis on division, separation, and the establishment of boundaries. God separates light from darkness (Genesis 1:4). He separates the waters above from the waters below (Genesis 1:6-7). He gathers the seas so that dry land appears (Genesis 1:9-10). He creates the sun and moon to separate day from night and to mark seasons and times

(Genesis 1:14–18). Each act of creation involves making a distinction. Each distinction generates a new dimension of order.²

In information theory, information is created precisely through the establishment of distinctions. A system that could be in any of many states contains no information until those possibilities are narrowed. When you distinguish one thing from another—light from darkness, land from sea—you create structure. You establish boundaries. You impose order on what was previously undifferentiated. The universe becomes intelligible because distinctions exist. Without distinctions, nothing could be named, measured, or known. Genesis portrays creation precisely as this: a sequence of distinctions, structured intentionally and purposefully. The theological significance is profound. God does not create by imposing arbitrary divisions on chaos. He creates by establishing meaningful order. The separations in Genesis are not random. They serve purposes. Light allows visibility. The separation of waters creates space for atmosphere and land. Day and night establish rhythms for life. The act of separating is the act of making the world habitable, knowable, and good. This is creation as the gift of intelligible structure, not merely the imposition of power.

The third feature is the recurring refrain: "And God said" (Genesis 1:3, 6, 9, 11, 14, 20, 24, 26). In Hebrew thought, speech is not merely sound but action. The Hebrew word **davar** means both "word" and "thing" or "event." A divine word is something that accomplishes what it expresses. God's speech is performative. He does not describe reality; He establishes it. When God says, "Let there be light," light comes into being. When God says, "Let the earth bring forth vegetation," vegetation appears. The Word is efficacious. It brings about what it declares.³

This understanding of divine speech connects deeply to the concept of information in physics. Physical interactions establish correlations, fix states, and determine outcomes. When a quantum system decoheres, information about its state becomes encoded in the environment. When particles interact, they exchange information that determines their subsequent behavior. The universe unfolds through the establishment of correlations and the actualization of potential states. In theological language, God speaks and reality responds. In physical language, interactions occur, and states become definite. Both describe a universe that comes into ordered form through processes that establish meaningful structure. The difference, of course, is that Scripture attributes this to personal agency. God intends creation to be ordered. He wills it to be good. He designs it to be fruitful. The universe is not a brute fact but the expression of divine purpose. Physics describes the mechanisms by which order emerges. Theology explains why there is order at all and why it serves the flourishing of life. One describes how distinctions function in nature. The other proclaims that those distinctions reflect the wisdom and generosity of the Creator.

The fourth feature is intelligibility and progression. Throughout the creation narrative, the world becomes increasingly ordered, structured, and capable of being known. Light reveals. Land supports life. Seasons mark time. The plants and animals fill their appointed places. The creation account moves from the simple to the complex, from the unformed to the patterned, from potential to fullness. This progression reflects the scientific understanding that the universe develops from simplicity toward complexity, from high symmetry toward broken symmetry, from undifferentiated energy toward structured form.[4]

You see this pattern everywhere in cosmology. The early universe was hot, dense, and nearly uniform. As it expanded and cooled, particles formed, forces differentiated, and structures emerged. Gravity pulled matter into clumps. Stars ignited. Galaxies assembled. Planets formed. Life appeared. Consciousness arose. At each stage, new levels of complexity emerged from simpler conditions. The progression was not random. It followed laws. Those laws permitted the development of richer forms of order. And that complexity eventually gave rise to conscious beings capable of understanding the laws themselves.

Genesis expresses this same trajectory in theological terms. God brings order from formlessness, distinction from undifferentiation, actuality from potentiality. Each day of creation builds upon the previous one. Light first, then atmosphere, then land and sea, then vegetation, then celestial bodies to govern time, then aquatic and aerial creatures, then land animals, and finally humanity made in God's image. The sequence is not arbitrary. It reflects a rational ordering, a progression toward greater complexity and toward the emergence of beings capable of knowing, naming, and relating to their Creator.

The fifth feature is teleology—the sense that creation has purpose and direction. Genesis does not portray the universe as an accident or a cosmic fluke. It portrays creation as good, ordered, and structured to support life. "And God saw that it was good" (Genesis 1:10, 12, 18, 21, 25, 31). This refrain echoes throughout the chapter. Creation is not merely functional. It is **good**. It reflects divine intention. It serves divine purposes.[5]

This echoes the fine-tuning discoveries of modern cosmology discussed earlier in this book. The physical constants of the universe—the strength of gravity, the mass of the electron, the balance between electromagnetic and nuclear forces—are precisely calibrated for the emergence of complexity, chemistry, stars, planets, and ultimately life itself. Change any of these values even slightly, and the universe would be sterile, incapable of producing galaxies, stars, or the chemical elements necessary for biology. The universe appears set up for intelligibility and habitation.

Genesis expresses this truth in its own idiom: God creates deliberately, not randomly. He intends the world to be ordered. He designs it to be fruitful. He structures it to support life. The scientific observation of fine-tuning and the theological declaration of purpose are not competing claims. They describe the same reality from different vantage points. One says, "The universe could easily have been inhospitable, but it is not." The other says, "God made the world to be inhabited, and it is good." Both affirm that reality is structured in a way that permits, and indeed fosters, the flourishing of life.

Genesis also emphasizes relationality. Light exists in relation to darkness. Land in relation to sea. Sun and moon in relation to one another. Plants in relation to the soil. Animals in relation to their environments. Human beings in relation to the world, to one another, and to God. Nothing in the Genesis account exists in isolation. Everything is part of a web of relationships, dependencies, and interactions.[6] This mirrors the relational nature of physical reality explored in earlier chapters. Quantum entanglement binds particles in immediate correlation regardless of distance. Spacetime geometry emerges from patterns of entanglement. The universe is relational at its most fundamental level. There are no truly isolated systems. Everything influences and is influenced by everything else. The cosmos is a network of interactions, not a collection of independent objects.

Genesis and modern physics thus agree on a profound truth: the universe is constituted by relationships, not by isolated substances. In theological language, God creates a world of interconnection, where creatures depend on one another and ultimately on Him. In physical language, the universe is a web of correlations, where no system exists independently of its environment. Both perspectives reject the notion of a mechanistic, atomistic cosmos in favor of a vision of reality as fundamentally relational and interconnected.

The climax of Genesis 1 is the creation of the human person. "Then God said, 'Let us make man in our image, after our likeness'" (Genesis 1:26, ESV). Humanity is made in the image of God—**imago Dei**. This does not refer to physical likeness but to rationality, agency, relationality, moral awareness, and the capacity for communion with the divine. Human beings are called to exercise stewardship, creativity, and understanding. They are invited into a relationship of knowledge and love with the Creator.[7]

This connects to the emergence of personhood explored earlier in this work. The universe unfolds in such a way that it produces beings capable of reflection, choice, and moral agency. Conscious creatures arise who can contemplate their own existence, investigate the laws of nature, create art and culture, and ask ultimate questions. This is not an accident. It is the culmination of a process that begins with the Big Bang and extends through billions of years of cosmic evolution. The universe, as Freeman Dyson observed, seems in some sense to have known we were coming.

Genesis expresses this in theological terms: God intends for the universe to produce persons who can know Him, love Him, and participate consciously in His purposes. Humanity is not an afterthought or a cosmic accident. We are the intended outcome of a creation ordered from the beginning toward the emergence of rational, relational beings. Our capacity to think, to choose, to love, to create—all of this reflects the image of God within us. And our longing for meaning, truth, goodness, and beauty is the echo of the divine calling us back to the Source from which we came.

The human vocation includes naming. Adam names the animals (Genesis 2:19–20). This is not trivial. To name something is to know it, to classify it, to understand its nature and place in the order of things. Naming is an act of intellectual mastery and stewardship. It reflects the human calling to understand creation, to discern patterns, to bring order to our knowledge of the world. Modern science continues this ancient task. When scientists name species, classify elements, map the genome, or discover new particles, they participate in the Adamic vocation of naming—of bringing the created order into the realm of human understanding.

Another key aspect of Genesis is that creation culminates in rest. God ceases from creating not because of exhaustion but because creation has reached a state of ordered goodness. "And on the seventh day God finished his work that he had done, and he rested on the seventh day from all his work that he had done" (Genesis 2:2, ESV). Rest signifies completeness. It marks the transition from formation to fulfillment. Creation is not merely the establishment of physical structures but the establishment of a meaningful order in which creatures can live, grow, and flourish.[8]

In a relational universe, Sabbath rest reflects the purpose of creation: communion, delight, and relationship. The universe is not a perpetual machine grinding through its paces. It is a home, a place where persons can dwell, where life can flourish, where love can be expressed and received. The Sabbath affirms that creation has a goal: not merely the production of more and more complexity, but the establishment of conditions under which God and His creatures can rest together, delighting in the goodness of what has been made.

It is crucial to emphasize that the purpose of this chapter is not to force scientific concepts into the biblical text or to bend theology to match physics. The creation account in Genesis is not a scientific treatise. It does not describe the mechanisms by which stars form or galaxies assemble. It does not provide equations or empirical predictions. Attempting to read Genesis as a physics textbook does violence to both Scripture and science.

Instead, the goal is to show that the **structure** of Genesis resonates with the deep structure of reality revealed by modern science. Both accounts affirm that the universe emerges through distinctions. Both portray creation as ordered, intelligible, and relational. Both regard the emergence of persons as the central development in the history of the cosmos. These are conceptual parallels, not mechanical correlations. They suggest that the ancient theological vision and the modern scientific picture are not enemies but complementary perspectives on the same reality.

Genesis speaks in the language of purpose, intention, and divine action. Physics speaks in the language of laws, mechanisms, and mathematical structure. Theology tells us **"why"** the universe exists and **"what it means"**. Science tells us **"how"** the universe behaves and **"what patterns"** it exhibits. Neither can replace the other. But when both are understood properly, within their own domains, they reveal a coherent vision: a universe that is rational, relational, ordered, and ultimately grounded in the creative Word of God.

Genesis 1, when read with care, describes creation as the articulation of meaningful distinctions through divine speech. It presents the universe as something that comes into being through order, intention, and relational structure. Light separated from darkness.

Waters above separated from waters below. Sea gathered so that land appears. Day distinguished from night. Each separation is an act of creation. Each distinction generates a new dimension of order. The formless void becomes a cosmos, a word that means both "order" and "beauty", because God speaks distinctions into existence.

Modern physics, from an entirely different starting point, arrives at a strikingly similar picture. The universe is not a static lump of matter but a dynamic, relational, informational reality. Structure emerges through the establishment of correlations. Order arises from the breaking of symmetries. Complexity develops from simplicity. And the entire process can be described, in part, using the language of information theory: the mathematics of distinctions, boundaries, and meaningful structure.

These themes harmonize naturally with the informational and relational universe described by physics. They prepare the way for the next chapter, which examines the Logos theology of the New Testament and shows how the Gospel of John unites the biblical depiction of creation with the philosophical idea of rational structure and divine order. If Genesis portrays God creating through speech, John takes this insight further: the Word through whom all things were made is not merely a principle but a person—the eternal Logos who became flesh and dwelt among us.

CHAPTER REVIEW:
KEY CONCEPTS:

- **Tohu wabohu** (formless void) Not absolute nothing but lack of functional order and distinction
- **Creation through separation** Light from darkness, waters above from below, sea from land; each separation generates order
- **Divine speech as creative act** "Let there be..." results in existence; Word is performative, not descriptive
- **Progressive actualization** From simple initial state to complex cosmos through sequential distinctions
- **Information-theoretic parallel** Information defined by

distinctions; Genesis describes creation establishing meaningful differences

WHAT THIS MEANS:

Genesis and information theory describe reality using the same fundamental concept: meaningful distinction generates order. This is not because Genesis anticipated modern physics, but because both are describing the actual structure of creation. One speaks theologically, the other mathematically, but they converge on a universe characterized by progressive differentiation and actualization.

IMPORTANT TERMS & CONCEPTS:

- **Tohu wabohu** → Chapter 13 endnotes; Part X theological terms
- **Davar** (word/event/thing) → Part X, Appendix B (Hebrew word studies)
- **Information as Distinction** → Part X, Section 1.4 (Information General Definition)
- **Creatio ex nihilo** → Part X, Section 1.1; Q&A Section II
- **Sabbath Rest** → Chapter 13 endnotes; theological significance

LOOKING AHEAD:

- Chapter 14 develops Logos theology showing Christ as the creative Word
- Chapter 15 connects Incarnation to creation's meaning
- This sets up Part V's treatment of the Triune God as Creator
- The informational framework developed here underlies later theological claims

TECHNICAL REFERENCE:

- For information vs. meaning distinction: Part X, Section 1.4 (complete definitions)
- For analogical use of "information" in theology: Part X, Section 7.5 (Analogical Interpretation)
- For what "Genesis as information" does NOT mean: Part X, Section 3
- For creation theology: Q&A Section II; Part X, Section 1.1 (Theology)

1. The Hebrew phrase **tohu wabohu** (וָבֹהוּ תֹהוּ) in Genesis 1:2 is often translated "without form and void" or "formless and empty." For linguistic and theological analysis: Tsumura, D. T. (1989). *The Earth and the Waters in Genesis 1 and 2: A Linguistic Investigation*. Sheffield Academic Press. Walton, J. H. (2001). *Genesis 1 as Ancient Cosmology*. Eisenbrauns. These works explore how ancient Near Eastern cosmology understood pre-creation states as lacking functional order rather than absolute nothingness, which aligns with the doctrine of **creatio ex nihilo** developed in later theology.
2. For the theological significance of separation and distinction in Genesis: Levenson, J. D. (1988). *Creation and the Persistence of Evil: The Jewish Drama of Divine Omnipotence*. Princeton University Press, Chapter 1. Kass, L. R. (2003). *The Beginning of Wisdom: Reading Genesis*. Free Press, Chapter 2. The act of separating establishes boundaries that make ordered existence possible. For the connection to information theory: Davies, P., & Gregersen, N. H. (Eds.). (2010). *Information and the Nature of Reality: From Physics to Metaphysics*. Cambridge University Press, especially Chapter 2 on distinctions and information.
3. The Hebrew concept of **davar** (word/event/thing) emphasizes the performative nature of divine speech. For theological treatment: Barr, J. (1961). *The Semantics of Biblical Language*. Oxford University Press, pp. 129–140. Wolterstorff, N. (1995). *Divine Discourse: Philosophical Reflections on the Claim that God Speaks*. Cambridge University Press. The scriptural basis is clear: "By the word of the LORD the heavens were made" (Psalm 33:6); "So shall my word be that goes out from my mouth; it shall not return to me empty, but it shall accomplish that which I purpose" (Isaiah 55:11, ESV).
4. For the progression from simplicity to complexity in cosmology: Silk, J. (2001). *The Big Bang*, 3rd ed. W. H. Freeman, Chapters 6–8. Weinberg, S. (1977). *The First Three Minutes: A Modern View of the Origin of the Universe*. Basic Books. For broken symmetry and structure formation: Guth, A. H. (1997). *The Inflationary Universe: The Quest for a New Theory of Cosmic Origins*. Perseus Books, Chapter 11. The universe begins in a highly symmetric, simple state and evolves toward increasing structural complexity.
5. The concept of creation as "good" in Genesis has profound theological implications for understanding divine purpose and design. For theological commentary: Barth, K. (1958). *Church Dogmatics III/1: The Doctrine of Creation*. T&T Clark, §41–42. Middleton, J. R. (2005). *The Liberating Image: The Imago Dei in Genesis 1*. Brazos Press, Chapter 2. For the connection to fine-tuning and teleology: Collins, R. (2009). "The Teleological Argument: An Exploration of the Fine-Tuning of the Universe." In Craig, W. L., & Moreland, J. P. (Eds.), *The Blackwell Companion to Natural Theology*. Wiley-Blackwell, pp. 202–281.
6. For the relational nature of reality in biblical theology: Zizioulas, J. D. (1985). *Being as Communion: Studies in Personhood and the Church*. St Vladimir's Seminary Press. LaCugna, C. M. (1991). *God For Us: The Trinity and Christian Life*. HarperSanFrancisco. For quantum relationality and entanglement: Rovelli, C. (1996). "Relational Quantum Mechanics." *International Journal of Theoretical Physics* 35(8), 1637–1678. The universe at the quantum level is constituted by rela-

tionships rather than independent substances, a view that resonates with biblical relationality.

7. The **imago Dei** doctrine is central to Christian anthropology. For comprehensive treatments: Middleton, J. R. (2005). *The Liberating Image: The Imago Dei in Genesis 1*. Brazos Press. McFarland, I. A. (2010). *The Divine Image: Envisioning the Invisible God*. Fortress Press. For the naming vocation: Anderson, B. W. (1977). "Human Dominion Over Nature." In *Biblical Studies in Contemporary Thought*. Somerville, MA: Greene, Hadden & Co., pp. 27–45. Adam's naming of the animals represents humanity's calling to understand and order creation.

8. For the theology of Sabbath and rest: Brueggemann, W. (2014). *Sabbath as Resistance: Saying No to the Culture of Now*. Westminster John Knox Press. Heschel, A. J. (1951). *The Sabbath: Its Meaning for Modern Man*. Farrar, Straus and Giroux. Sabbath rest signifies that creation is ordered not merely toward production but toward relationship, delight, and communion with God. For the culmination of creation in personhood: Polkinghorne, J. (2005). *Exploring Reality: The Intertwining of Science and Religion*. Yale University Press, Chapter 6.

CHAPTER FOURTEEN
THE LOGOS OF JOHN 1 AND THE FOUNDATIONS OF REALITY

The opening of the Gospel of John stands among the most influential sentences ever written. "In the beginning was the Word, and the Word was with God, and the Word was God" (John 1:1, NABRE). With these words, the evangelist unites the Hebrew account of creation with the philosophical quest of the Greek world. He identifies the Logos as both divine and personal, eternal and active, the One through whom all things came into existence. This chapter explores the meaning of the Logos in its biblical, philosophical, and cosmological dimensions, and shows how the concept of the divine Word resonates with the ordered structure of the universe.

The first thing you need to understand is that John does not invent the word **Logos**. He chooses a term already filled with centuries of meaning, a word that carried profound significance in both Jewish and Greek thought. In the Hebrew Scriptures, God creates through His word. "By the word of the LORD the heavens were made, and by the breath of his mouth all their host" (Psalm 33:6, ESV). God speaks, and creation responds. Divine speech is not merely communication but creative action. It establishes order, gives identity, and calls new realities into being. The Hebrew word

davar conveys both meaning and event. A word accomplishes what it expresses. When God says, "Let there be light," light comes into existence. The word is not a description of what happens; it is the cause of what happens.[1]

Greek philosophy also contributed to the concept of Logos, though from a different angle. For Heraclitus, writing in the sixth century BC, Logos was the rational principle that permeates the universe: the hidden order behind the apparent chaos of change. Fire might consume wood, water might freeze into ice, day might turn to night, but underlying all these transformations was a stable pattern, a rational structure that gave coherence to the cosmos. The Stoics developed this idea further. For them, Logos was the divine rationality immanent in nature, the principle of coherence and the source of natural law. They believed that human rationality was a spark of the universal Logos, which meant that the human mind could grasp the structure of reality precisely because both mind and reality shared a common rational foundation. In Greek thought, Logos united reason, structure, and purpose.[2]

John draws these traditions together and does something unprecedented. He identifies the Logos not merely as divine reason or divine speech, but as a divine person. "The Word was with God, and the Word was God" (John 1:1, ESV). The Logos is not an abstract principle. The Logos is not an impersonal force. The Logos is eternal, personal, and fully divine. And then comes the staggering claim: "All things were made through him, and without him was not anything made that was made" (John 1:3, ESV). In this single declaration, the evangelist provides a theological interpretation of the universe's intelligibility. The world is intelligible because it comes from the divine Logos. The universe is ordered because it is spoken into existence by the One who is perfect rationality.

This understanding aligns deeply with what modern physics reveals. The universe is not a chaos of unconnected events. It is lawful, structured, and rationally intelligible. Symmetry principles, mathematical regularities, and informational constraints govern everything from subatomic particles to galaxies. This profound order has long puzzled scientists and philosophers alike. Why should nature obey mathematics so readily? Why should the universe be structured in a way that minds can understand? Eugene Wigner famously called this "the unreasonable effectiveness of mathematics in the natural sciences". Unreasonable because there is no obvious necessity for it, yet undeniable because it is the foundation of all scientific progress.³

The doctrine of the Logos provides an answer. Creation is rational because it flows from divine rationality. The laws of nature are not arbitrary conventions imposed on a recalcitrant universe. They are expressions of the Logos. The universe is intelligible because the Logos is intelligible. Human intelligence can grasp the world because human beings are made in the image of the Logos. The rationality of nature is not an accident or a lucky coincidence. It reflects the character of the One who made it. When you solve an equation that describes the motion of planets or the behavior of quantum fields, you are not imposing human categories on nature. You are discovering patterns that were there from the beginning, patterns that reflect the rationality of the One through whom all things were made.

John goes even further. "In him was life, and the life was the light of men" (John 1:4, ESV). The Logos is not only the source of physical order but the source of life and illumination. Life itself is an expression of the Logos. Consciousness is an echo of the divine intelligibility. Human rationality is a participation in the eternal Word. This understanding harmonizes with the informational view of consciousness explored earlier in this book. If consciousness arises from the integration of information. If awareness depends on structured patterns of relationships within the brain, then it mirrors the divine Wisdom through which the universe itself is structured. You think, you understand, you know because the universe is fundamentally knowable, and the universe is knowable because it comes from the Logos.

The most striking claim in John's prologue is what happens next: "And the Word became flesh and dwelt among us, and we have seen his glory, glory as of the only Son from the Father, full of grace and truth" (John 1:14, ESV). The eternal Logos enters human history. The Creator enters creation. The rational ground of being becomes a particular human being, living in first-century Palestine, walking on dusty roads, teaching in synagogues, eating meals with tax collectors and sinners. This claim is central to Christian theology. It reveals the personal nature of God, the relational depth of the divine, and the purpose of creation. The Logos does not remain distant, aloof, transcendent in some unreachable heaven. He becomes present in the world He made. He bridges the infinite gap between the uncreated and the created.[4]

From the perspective of the (IAM), this moment has profound implications. If the universe is the unfolding of divine meaning. If creation is the articulation of divine purpose through ordered distinctions and relational structure, then the Incarnation is the point at which the divine meaning enters the narrative from within. The Logos who creates through information, order, and relationality now enters creation as a person, revealing fully the nature of God and the dignity of human beings. The universe exists through the Word. The Word steps into the universe to reveal its purpose. And in doing so, He shows that the deepest reality of the cosmos is not matter, not energy, not even information in some abstract sense, but personal love expressed in relational communion.

Another important dimension of the Logos theology is its connection to Wisdom literature in the Hebrew Scriptures. In Proverbs, Wisdom is personified as being with God at the beginning of creation, delighting in the world and in humanity. "The LORD possessed me at the beginning of his work, the first of his acts of old. Ages ago I was set up, at the first, before the beginning of the earth... When he established the heavens, I was there... then I was beside him, like a master workman, and I was daily his delight, rejoicing before him always" (Proverbs 8:22–23, 27, 30, ESV). Wisdom is portrayed as the pattern through which creation is formed, the order through which chaos becomes cosmos.[5]

This portrayal closely resembles the philosophical and scientific concept of the universe's deep structure. The Logos of John fulfills the role of Wisdom, revealing that the structure of the universe is not an impersonal pattern but a personal reality. The order you observe in nature—the elegance of physical laws, the beauty of mathematical symmetry, the fine-tuning that makes life possible—all of this reflects the Wisdom of God expressed through the Logos. When you study science, you are not merely cataloging facts. You are tracing the thoughts of God as they are inscribed in the fabric of creation.

The Logos also connects to the moral order. The Stoics understood Logos as the source of natural law, the rational structure that grounds right action. They believed that living in accordance with nature meant living in accordance with the Logos, aligning one's will with the rational structure of reality. John's theology deepens this insight. The moral law is not an abstract principle floating somewhere above the universe. It is an expression of the character of the Logos. Human morality reflects participation in divine rationality. To act in truth is to act in harmony with the Logos. To distort truth is to act against the structure of reality itself. When you lie, you do not merely violate a social convention. You contradict the nature of the One who is Truth. When you act justly, you do not merely follow rules. You participate in the justice that flows from the character of God.[6]

One more aspect deserves attention. The Logos is not only the source of creation but its sustainer. In the language of Christian theology, developed especially in Colossians and Hebrews, the Logos continually holds all things in being. "He is before all things, and in him all things hold together" (Colossians 1:17, NABRE). This aligns with the metaphysical understanding discussed in earlier chapters: God's act of being is the ground of all existence. The universe does not persist on its own. It does not wind itself up at the beginning and then run on its own like a clock. It exists because God continuously wills it to exist.

Every moment of the universe's existence is sustained by the eternal act of the Logos. When Paul writes that "in him all things hold together" (συνέστηκεν - literally "cohere," "stand together"), he's making a cosmological claim about Christ's ongoing sustaining role. Modern physics has discovered something remarkable: the universe is dominated (~68%) by a mysterious "dark energy", represented by Einstein's cosmological constant Λ, that prevents gravitational collapse and maintains cosmic structure. This isn't mere analogy. Both describe the same fundamental reality:

- A sustaining force present everywhere
- Preventing collapse into nothingness
- Invisible but structurally essential
- The reason existence continues rather than ceases

The convergence is even more striking when we recognize that this cosmological constant is represented by Λ (Lambda)—the first letter of Λόγος (Logos), the divine Word through whom "all things were made" (John 1:3) and by whom they are sustained.

What physics calls "the force preventing cosmic collapse" and what theology calls "the Word in whom all things hold together" are not competing explanations but complementary descriptions at different levels of analysis. The informational unfolding of the cosmos—the way quantum states decohere, the way spacetime evolves, the way galaxies form and stars burn and life emerges—all of this depends on the eternal act by which the Logos shares being with what is not God.[7]

This has practical implications for how you understand your own existence. You are not a self-sufficient entity, a little island of being maintaining yourself by your own power. You exist at this moment because the Logos wills you to exist. Your consciousness, your rationality, your capacity for love and creativity—all of these are gifts sustained by the One through whom all things were made. To exist is to be held in being by the Logos. To know is to participate in the rationality of the Logos. To love is to reflect the relational nature of the One who is Love. Your life is not an accident. It is part of the ongoing creative and sustaining work of the Logos, who continually speaks the universe into existence.

The chapter has shown that the New Testament understanding of the Logos provides a profound account of why the universe is rational, relational, intelligible, and capable of producing persons. The Logos is the bridge between God and creation, between the infinite and the finite, between intelligibility and existence. The Logos is the ground of natural law, the source of consciousness, and the purpose of creation. The Logos theology of John unites Hebrew revelation and Greek philosophical insight in a single vision that resonates with the relational structure of the universe revealed by modern physics.

But John does not stop with the Logos as Creator and Sustainer. He insists that this same Logos "became flesh and dwelt among us." This moves us beyond cosmology and metaphysics into the realm of history, incarnation, and redemption. If the Logos is personal and becomes human, what does this reveal about the nature of God and the meaning of personhood? How does the Incarnation illuminate the relational basis of the universe and the dignity of every human life? The next chapter explores these questions by examining Christ as the embodied Logos—the One through whom the deepest truths of creation become visible in human form.

CHAPTER REVIEW:
KEY CONCEPTS:

- **Logos = divine Reason/Word through whom all things were made** Combining Greek philosophical concept with Hebrew davar

- **John 1 unites creation, reason, and incarnation** "In the beginning was the Word...all things were made through him"

- **Logos as the rational structure of reality** Universe is intelligible because it proceeds from the Logos

- **Personal rather than abstract** Logos is not a principle but a Person (Christ)

- **Bridge between infinite and finite** Logos mediates between transcendent God and created order

WHAT THIS MEANS:

The universe's mathematical intelligibility and rational order are not accidents but reflections of their source in the Logos. Physics discovers this rational structure; theology reveals its personal ground. The Logos doctrine shows why the universe makes sense to minds and why truth is one even when approached through different disciplines.

IMPORTANT TERMS & CONCEPTS:

- **Logos** → Part X, Section 1.1 (Theology); Appendix B (Greek word studies)
- **Heraclitus & Stoics** (Logos in philosophy) → Chapter 14 endnotes
- **Davar** (Hebrew word) → Appendix B
- **Intelligibility** → Part X, Section 1.1 (Philosophy domain)
- **Incarnation** → See Chapter 15; Part X, Q&A Section V

LOOKING AHEAD:

- Chapter 15 explores Christ as embodied Logos
- Part V develops Trinitarian theology as ultimate reality
- This foundation supports later claims about reason, truth, and revelation
- Q&A Section V addresses specific questions about Christ and Logos

TECHNICAL REFERENCE:

- For Logos properly understood: Part X, Section 1.1 (Theology); Appendix B
- For why Logos ≠ information: Part X, Section 7.5 (Analogical Interpretation)
- For intelligibility of universe: Part X, Section 1.1 (Philosophy domain)
- For Christology: Q&A Section V (Questions 41-50)

---

1. For the Hebrew concept of **davar** as both word and event: Boman, T. (1960). *Hebrew Thought Compared with Greek.* SCM Press, pp. 58–69. Theological treatment of divine creative speech: Wolterstorff, N. (1995). *Divine Discourse: Philosophical Reflections on the Claim that God Speaks.* Cambridge University Press. The performative nature of God's word is emphasized throughout Scripture:

"For he spoke, and it came to be; he commanded, and it stood firm" (Psalm 33:9, ESV); "So shall my word be that goes out from my mouth; it shall not return to me empty, but it shall accomplish that which I purpose" (Isaiah 55:11, ESV).

2. For Heraclitus and the concept of Logos: Kirk, G. S., Raven, J. E., & Schofield, M. (1983). *The Presocratic Philosophers*, 2nd ed. Cambridge University Press, Chapter 6. For Stoic development of Logos: Long, A. A. (1986). *Hellenistic Philosophy: Stoics, Epicureans, Sceptics*, 2nd ed. University of California Press, Chapter 5. The Stoics taught that human reason (**logos endiathetos**) is a fragment of the universal Logos (**logos spermatikos**) pervading all nature.

3. Wigner, E. (1960). "The Unreasonable Effectiveness of Mathematics in the Natural Sciences." *Communications in Pure and Applied Mathematics* 13(1), 1–14. For philosophical treatment of mathematics and physical law: Steiner, M. (1998). *The Applicability of Mathematics as a Philosophical Problem*. Harvard University Press. Barrow, J. D. (1992). *Pi in the Sky: Counting, Thinking, and Being*. Oxford University Press. The intelligibility of nature is not self-explanatory and requires metaphysical grounding.

4. For theological treatment of the Incarnation and John's prologue: Brown, R. E. (1966). *The Gospel According to John (I–XII)*. Anchor Bible Commentary. Doubleday, pp. 3–37. Keener, C. S. (2003). *The Gospel of John: A Commentary*. Baker Academic, Vol. 1, pp. 333–425. The claim that "the Word became flesh" (**ho logos sarx egeneto**) is unique in ancient literature and central to Christian orthodoxy. For its metaphysical implications: Torrance, T. F. (1981).*Divine and Contingent Order*. Oxford University Press.

5. For the connection between Wisdom literature and Logos theology: Dunn, J. D. G. (1989). *Christology in the Making*, 2nd ed. Eerdmans, Chapter 6. Hengel, M. (1974). *Judaism and Hellenism*. Fortress Press, Vol. 1, pp. 153–175. Proverbs 8:22–31 portrays Wisdom (**Chokhmah** in Hebrew, **Sophia** in Greek) as present with God before creation and delighting in His works. Early Christian theology identifies Christ as the Wisdom of God: "Christ the power of God and the wisdom of God" (1 Corinthians 1:24, ESV).

6. For natural law, Stoic ethics, and Christian moral theology: Inwood, B., & Gerson, L. P. (Eds.). (2008). *The Stoics Reader: Selected Writings and Testimonia*. Hackett, pp. 185–226. For Christian development: Aquinas, T. *Summa Theologica*, I-II, Questions 90–94 (on natural law). Porter, J. (1999). *Natural and Divine Law: Reclaiming the Tradition for Christian Ethics*. Eerdmans. The moral structure of reality reflects the character of the Logos, not arbitrary divine commands.

7. For the sustaining work of the Logos: Colossians 1:15–17, Hebrews 1:1–3. Theological treatment: Gunton, C. E. (1998). *The Triune Creator: A Historical and Systematic Study*. Eerdmans, Chapter 8. Torrance, T. F. (1996). *The Christian Doctrine of God, One Being Three Persons*. T&T Clark, pp. 207–220. Classical doctrine of **creatio continua** (continuous creation) affirms that God not only brings the universe into being but sustains it in existence at every moment. This is distinct from deism, which posits a distant God who creates but does not sustain.

CHAPTER FIFTEEN
CHRIST AS THE EMBODIED LOGOS

*I*f the Logos is the rational, relational, life-giving source of all creation, then the Christian claim that "the Word became flesh" (John 1:14, ESV) carries immense philosophical and theological significance. It means that the foundational principle of the universe is not an abstraction, not a force, not a distant intellect, but a person. It means that the One through whom all things were made entered the world He created. It means that the universe's deepest truth is personal.

This chapter examines the meaning of the Incarnation in light of the informational, relational universe explored throughout this book. The goal is not to reduce theology to physics, but to show how the Christian understanding of Christ as the embodied Logos illuminates the nature of reality and the purpose of human existence.

At the heart of the Incarnation is the claim that the eternal Son, the second person of the Trinity, truly became human. He did not merely appear human, as some ancient heresies suggested. He did not inhabit a human body temporarily, like a spirit possessing a vessel. He assumed human nature in its fullness: body and soul, matter and mind, rationality and emotion. In doing so, He united the divine and human in one person without confusion, without change, without division, without separation. This is the doctrine of the hypostatic union, carefully articulated at the Council of Chalcedon in AD 451, and it remains the heart of orthodox Christian faith.[1]

This union reveals something profound about the universe itself. If God can enter creation as a human being, then creation is not alien to God. It is not a realm of chaos fundamentally opposed to the divine. It is not, as the Gnostics imagined, a prison from which we must escape. It is a domain shaped by divine wisdom, a world capable of receiving God's presence. The intelligibility and relationality of the universe reflect this compatibility. The world is structured in such a way that the Creator can enter it without contradiction. The laws of physics do not need to be suspended for God to become human. The structure of matter does not need to be violated. The Logos who established those laws and that structure can work within them, fulfilling them rather than breaking them.

This tells you something essential: the physical world is good. Matter is good. The body is good. These are not obstacles to the divine but vehicles for divine presence. When the Logos takes on flesh, He affirms the goodness of creation. He sanctifies the material order by entering it. And this has implications for how you understand your own embodied existence. You are not a ghost trapped in a machine. You are not a soul imprisoned in flesh. You are a unified person, body and soul, made in the image of God and called to live as an embodied creature in a physical world that God Himself has honored by entering it.[2]

The Incarnation also reveals the personal nature of the foundation of reality. The Logos is not simply rational structure or informational order, as if the universe were grounded in a principle or a formula. He is a person. Rationality, meaning, and purpose are grounded in personal being. The universe is intelligible because it comes from One who is intelligible. It is relational because it comes from One who is relational. It is capable of producing persons because it is created by a personal God. When modern physics describes the universe in terms of information, correlations, and relational structure, it is describing something true about reality. But the ultimate explanation for why reality has this structure is not found in the equations themselves. It is found in the Person who is the source of those equations, the Logos through whom all things were made.[3]

The Incarnation shows that human nature has inherent dignity. If the Logos assumes human nature, then human nature is not an accident or a byproduct of blind evolutionary forces. It is part of God's eternal plan. The rationality, creativity, moral awareness, freedom, and capacity for love that define human beings are not evolutionary anomalies that happened to emerge in one species on one planet. They reflect the nature of the One who made us. When Christ takes on human nature, He affirms that humanity is meant for communion with God. He elevates human nature to a status beyond anything imaginable: union with the divine. And if God became human, then every human life—from conception to natural death, in sickness and in health, in strength and in weakness—possesses infinite dignity. To harm a human being is to attack the nature that God Himself assumed. To despise the weak is to despise the One who became weak for us.[4]

The Incarnation reveals the purpose of creation. The Logos enters the world not only to reveal God but to heal and restore creation. He becomes human in order to reconcile humanity to God, to redeem what is fractured, and to elevate human nature to participation in the divine life. This purpose is woven into the fabric of creation itself. The universe is not merely a physical system running through its paces. It is a story, a narrative shaped by divine love. And Christ stands at the center of that story. The arc of cosmic history bends toward the Incarnation. Everything before it is preparation. Everything after it is response. The universe exists so that the Logos might become flesh and dwell among us.

The Incarnation also sheds light on the nature of knowledge. If the Logos is the source of all truth, then to know anything truly is, in a sense, to know something about Him. Mathematics describes patterns that reflect the rationality of the Logos. Physics discovers laws that express His wisdom. Moral truth reveals His character. Personal relationships mirror the Trinitarian communion from which all relationality flows. The Incarnation unites these domains. Christ is the bridge between the intellectual order of the universe and the personal order of salvation. Truth is not merely conceptual; it is personal. When Pilate asked, "What is truth?" (John 18:38, ESV), he was standing before Truth incarnate. And the answer was not a proposition but a Person.[5]

Another aspect of the Incarnation is its revelation of relationality. Christ's life demonstrates that the deepest reality is not isolation but communion. The relationship between Father, Son, and Spirit is the eternal source of all relational structure in the universe. God is not a solitary monad who decides, for inscrutable reasons, to create something outside Himself. God is three persons in perfect communion, eternally giving and receiving love. And when the Son becomes human, He extends this divine relationality into creation. He lives in relationship with others. He calls disciples. He teaches love. He forgives enemies. He establishes a community. His life embodies the relational nature of being. And this has implications for how you understand yourself. You are not made to be alone. You are not a self-sufficient individual who needs no one. You are made for relationship, with God and with others. Your deepest fulfillment comes not from autonomy but from communion.[6]

The Incarnation also illuminates the meaning of suffering. If the Logos enters a world marked by imperfection, risk, and moral failure, He does so knowingly. He does not exempt Himself from the conditions of human existence. He experiences human limitation. He feels sorrow and pain. He weeps at the tomb of Lazarus. He agonizes in Gethsemane. He endures the cross. This reveals that God is not distant from creation's struggles. He enters them. He shares them. He transforms them from within. The informational unfolding of the universe contains the possibility of suffering because it contains the possibility of freedom. Freedom requires openness. Openness entails risk. Risk includes the possibility of harm. Christ's suffering reveals that God respects this freedom and works within it rather than removing it. He does not wave a hand and make all pain vanish. He enters the pain and redeems it.[7]

Furthermore, the Incarnation reveals that the universe is oriented toward glory. Christ's resurrection and ascension show that human nature is destined for transformation. The resurrection is not merely the reanimation of a corpse. It is the glorification of human nature, the transformation of the physical body into something incorruptible, immortal, and radiant with divine life. And what happens to Christ in the resurrection is the prototype for what will happen to all who are united to Him. "We shall be like him, because we shall see him as he is" (1 John 3:2, ESV). The ordered and relational structure of the universe is not complete. It points toward a future fulfillment. The Incarnation is the beginning of that fulfillment. Through Christ, the universe's story moves toward restoration, renewal, and the full revelation of God's presence.[8]

Finally, the Incarnation shows that the ground of being is not merely the source of existence but the source of love. The eternal I AM enters the world not to display raw power but to draw humanity into communion. The Logos becomes flesh because God desires relationship. Creation begins in generosity and is completed in love. The Incarnation is the definitive expression of this divine intention. And this transforms how you understand your own existence. You are not a cosmic accident. You are not the product of random forces in an indifferent universe. You are loved by the One who made you, known by the One who sustains you, called by the One who became human for you. The invitation to communion with God is not an afterthought. It is the reason the universe exists.

This chapter has shown that Christ as the embodied Logos is the key to understanding the purpose of creation, the dignity of human nature, the personal foundation of reality, and the relational structure of the universe. The Incarnation unites physics, metaphysics, and theology in a single vision: the universe is grounded in personal being, created through rational order, shaped by relationality, and directed toward communion. The Word through whom all things were made became flesh to reveal God fully and to bring creation to its intended fulfillment. In Christ, you see not only who God is but also who you are meant to become.

The next section of this book turns from the Logos to the nature of God as revealed in Scripture. If Christ reveals the personal foundation of reality, what does the divine name "I AM That I AM" teach about the ontology of God? And how does this understanding shape your view of existence, personhood, and the destiny of creation? The connection between the Logos and the divine name is not incidental. It is essential. For the One who revealed Himself to Moses as "I AM" is the same One who revealed Himself in Christ as the Word made flesh.

CHAPTER REVIEW:
KEY CONCEPTS:
- **Incarnation: Logos becomes flesh** John 1:14; the divine Reason takes on human nature
- **Imago Dei** Humans as image-bearers participate in the Logos through rationality and relationality
- **Christ reveals both God and true humanity** Shows what God is like AND what humans are meant to be
- **Redemption restores right relationship** Incarnation addresses the fracture between Creator and creation
- **Personal nature of ultimate reality** Being Itself is not abstract but relational and personal

WHAT THIS MEANS:
The Incarnation reveals that ultimate reality is personal, not just mechanical or abstract. The Logos through whom all was made takes on flesh, showing that matter matters and that the physical universe is the arena of divine action. This grounds human dignity (image of God) and shows that redemption is about restored relationship, not escape from physicality.

IMPORTANT TERMS & CONCEPTS:
- Incarnation → Part X, Q&A Section V; theological terms

- Imago Dei (Image of God) → Part X, Q&A Section III; Chapter 22
- Hypostatic Union → Part X, Q&A Section V (Christ's two natures)
- Redemption → Part X, Q&A Section VI
- Person (theological sense) → Part X, Section 1.6; See Part V (Trinity)

LOOKING AHEAD:
- Part V develops Trinitarian theology showing relationality at the heart of reality
- Chapter 22 explores what it means to be known by I AM
- Q&A Section V addresses specific Christological questions
- This grounds the later treatment of human dignity and purpose

TECHNICAL REFERENCE:
- For Incarnation theology: Part X, Q&A Section V (Questions 41-50)
- For imago Dei: Q&A Section III (Questions 21-30)
- For person vs. nature distinction: Part X, Section 1.1 (Theology)
- For what embodiment means theologically: Q&A Sections III & V

---

1. The Council of Chalcedon (AD 451) defined the doctrine of the hypostatic union: Christ is "acknowledged in two natures, without confusion, without change, without division, without separation." For historical context: McGuckin, J. A. (2004). *The Westminster Handbook to Patristic Theology*. Westminster John Knox Press, pp. 101-106. For theological treatment: Torrance, T. F. (1992). *The Trinitarian Faith: The Evangelical Theology of the Ancient Catholic Church*. T&T Clark, pp. 147-180. The Incarnation involves a real union of divine and human natures in one person.
2. For the goodness of creation and matter in Christian theology: Genesis 1:31 ("And God saw everything that he had made, and behold, it was very good," ESV). Theological development: Gunton, C. E. (1998). *The Triune Creator: A Historical and Systematic Study*. Eerdmans, Chapter 4. The Incarnation affirms that matter is not evil but capable of bearing divine presence. Contrast with Gnostic dualism: Jonas, H. (1958). *The Gnostic Religion*, 2nd ed. Beacon Press.
3. For the personal nature of ultimate reality and the Logos: Zizioulas, J. D. (1985). *Being as Communion: Studies in Personhood and the Church*. St Vladimir's Semi-

nary Press. LaCugna, C. M. (1991). *God For Us: The Trinity and Christian Life*. HarperSanFrancisco. Being itself is personal and relational, not impersonal and mechanical. The universe's structure reflects the personal rationality of its Creator.
4. For the dignity of human nature affirmed by the Incarnation: Athanasius, *On the Incarnation* (4th century), especially Sections 8–10. Modern treatment: Middleton, J. R. (2005). *The Liberating Image: The Imago Dei in Genesis 1*. Brazos Press, Chapter 1. O'Donovan, O. (1994). *Resurrection and Moral Order*, 2nd ed. Eerdmans, Chapter 2. The assumption of human nature by the Logos establishes humanity's permanent dignity and destiny.
5. For Christ as Truth incarnate and the unity of knowledge: Augustine, *Confessions*, Book VII. Barth, K. (1956). *Church Dogmatics II/1: The Doctrine of God*. T&T Clark, §25–26. All truth is God's truth, and Christ is the Logos through whom all truth is grounded. For the question of Pilate: see John 18:37–38 in context.
6. For the Trinity as the foundation of relationality: Zizioulas, J. D. (1985). *Being as Communion*. St Vladimir's Seminary Press. Volf, M. (1998). *After Our Likeness: The Church as the Image of the Trinity*. Eerdmans. Human personhood is fundamentally relational, reflecting the eternal communion of Father, Son, and Spirit. Isolation is contrary to the nature of being.
7. For the theology of divine suffering and the Incarnation: Moltmann, J. (1974). *The Crucified God: The Cross of Christ as the Foundation and Criticism of Christian Theology*. SCM Press. Weinandy, T. G. (2000). *Does God Suffer?* University of Notre Dame Press. While classical theism affirms divine impassibility, the Incarnation reveals that God enters human suffering in the person of Christ. He does not remain untouched by creation's pain.
8. For the resurrection as transformation and glorification: Wright, N. T. (2003). *The Resurrection of the Son of God*. Fortress Press, especially Part VI. 1 Corinthians 15:35–58 (on the resurrection body). The resurrection is not mere resuscitation but the transformation of human nature into glorified, incorruptible form. This is the destiny toward which all creation moves: Romans 8:18–25.

PART FIVE
THE RELATIONAL GOD

CHAPTER SIXTEEN
"I AM THAT I AM": GOD'S NAME AND ONTOLOGY

Among all the revelations contained in Scripture, none is more profound than the divine name given to Moses at the burning bush. When Moses asks who he should say has sent him to deliver Israel from Egypt, God responds with words that have echoed through theology and philosophy for millennia: "God said to Moses, 'I AM WHO I AM.' And he said, 'Say this to the people of Israel: "I AM has sent me to you"'" (Exodus 3:14, ESV). This answer stands at the center of biblical theology, classical metaphysics, and the (IAM) . It tells you not only who God is but what it means for anything to exist at all.

The Hebrew phrase is **Ehyeh Asher Ehyeh** (אֶהְיֶה אֲשֶׁר אֶהְיֶה). The verb **ehyeh** is the first-person singular imperfect form of **hayah**, meaning "to be" or "to exist." The grammatical structure resists simple translation. Depending on how you understand the Hebrew verb forms and the function of **asher** (which can mean "who," "that," "what," or serve as a relative pronoun), the phrase can legitimately be rendered in several ways: "I AM WHO I AM," "I AM WHAT I AM," "I AM THAT I AM," "I WILL BE WHO I WILL BE," or even "I WILL BE WHAT I WILL BE." The ancient Greek translation known as the *Septuagint* chose **ego eimi ho on** (ἐγώ εἰμι ὁ ὤν), typically translated as "I AM THE ONE WHO IS" or simply "I AM HE WHO IS."[1]

This multiplicity of possible translations is not a defect. It reveals the richness of the divine name. Each rendering captures something true about God's nature. "I AM WHO I AM" emphasizes God's self-existence and self-definition, He is not defined by anything outside Himself. "I AM WHAT I AM" stresses that God's essence and existence are identical. "I WILL BE WHO I WILL BE" highlights God's sovereign freedom and His faithfulness to fulfill His promises. "I AM HE WHO IS" (the *Septuagint* rendering) became the foundation for centuries of Christian metaphysics, particularly in the work of the Church Fathers and Thomas Aquinas, who saw in this name the clearest biblical statement that God is Being Itself.[2] This chapter examines the divine name as a statement of ontology, identity, and presence. It shows how the biblical revelation aligns with the philosophical understanding of God as Being Itself, and how this understanding illuminates the structure of the universe.

The first thing you need to recognize is that the divine name is utterly unique in the ancient world. The gods of surrounding cultures were identified by their roles, powers, or domains. Baal was the storm god. Molech was associated with fire and sacrifice. Marduk was the patron deity of Babylon. These were beings within the world, powerful perhaps, but finite and localized. They had genealogies, conflicts, limitations. They were characters in mythological narratives, subject to fate or chance or one another's machinations. The God of Israel reveals Himself in an entirely different category. He does not say, "I AM the God of storms" or "I AM the God who dwells on Mount Sinai." He says, "I AM." The revelation is ontological before it is anything else. God identifies Himself not as one being among many but as the source of all being.[3] "I AM" signifies absolute existence, perfect actuality, and complete self-sufficiency. God does not receive being from another. God does not participate in being as creatures do. God **is** being. Classical theology describes this reality using technical terms that have become foundational to Western metaphysics: divine **simplicity**, meaning God has no parts or composition; and divine **aseity** (from the Latin **a se**, meaning "from Himself"), meaning God exists independently, deriving His existence from no source outside Himself. God is not dependent on anything. He does not change, grow, develop, or decline. God's essence is His existence. In every other thing, essence and existence are distinct. A tree is one kind of thing and happens to exist, a rock is another kind of thing and happens to exist—but in God alone, to be God and to exist are the same thing. He simply is.[4]

This does not mean that God is static or inert, as if divine perfection implied some kind of frozen, lifeless immobility. Rather, God's life is the fullness of actuality, the infinite richness of being without limitation or potentiality. God is pure act (**actus purus** in the scholastic tradition), meaning there is nothing in God that is merely potential, nothing that could be but isn't yet. Everything else exists because God wills it to exist. Everything else participates in being, receiving existence moment by moment as a gift. Only God possesses being by nature, essentially, necessarily. The divine name also reveals God's presence. The Hebrew verb form **ehyeh** carries a sense of active, continuous, dynamic existence. This is not the "being" of a static concept or a timeless abstraction. It is the being of the living God who acts, who hears, who speaks, who delivers. The context of Exodus 3 makes this clear. Moses encounters God in a burning bush that is not consumed—a powerful symbol of divine holiness, transcendence, and inexhaustible life. God reveals Himself not from some distant heaven but from within the world, at a specific place, to a specific person, for a specific purpose. The One who is I AM hears the cry of His oppressed people. He remembers His covenant. He acts in history. The burning bush is not only a symbol of holiness but a sign of God's nearness.[5]

Being Itself, then, is not impersonal. It is the living God—the God who calls Moses by name, who commissions him, who promises to be with him. The God who is pure actuality, whose essence is existence, is also the God who enters into covenant relationship with humanity. This is the profound mystery at the heart of biblical theology: the God who transcends all creation is also the God who is personally present within it. The infinite is intimately near. The absolute is relationally engaged. This understanding aligns remarkably with the metaphysical structure of the universe explored throughout this book. The universe does not contain within itself the reason for its own existence. It points beyond itself. Every contingent being—every star, every atom, every quantum field—depends on something else for its existence. The chain of dependency cannot regress infinitely, as that would explain nothing. It must terminate in a being whose existence is not derived, whose essence includes existence, who is necessary rather than contingent. The divine name identifies this being with unmistakable clarity. God is the One whose essence is existence, the necessary ground upon which all contingent things depend.[6]

The divine name also sheds light on the nature of the universe. If God is Being Itself, then creation is not the emanation of a lesser deity working with pre-existing materials. It is not the product of chaos or accident. It is not the overflow of divine essence by necessity, as some ancient philosophies imagined. It is the free act of the One who is fullness of being. God creates not because He needs anything, He lacks nothing, but because His nature is generosity. Creation reflects the rationality, order, and relationality of its source. The intelligibility of the universe is grounded in the intelligibility of God. The relational structure of quantum mechanics, the informational basis of physical law, the fine-tuning that permits life: all of this points back to the Logos through whom all things were made, the Word spoken by the One who is I AM. The divine name also clarifies the distinction between God and creation, preventing two errors that have plagued theology and philosophy throughout history. The first error is **materialism**, which denies the existence of anything beyond the physical world, reducing all reality to matter and its interactions. But the physical world cannot explain its own being. It is contingent. It changes. It depends. It points beyond itself to a ground of being that is not physical, not material, not contingent. The divine name declares that this ground is not merely another physical thing but the source of existence itself.

The second error is **pantheism**, which identifies the world with God, dissolving the distinction between Creator and creation. In pantheism, everything is God and God is everything—the universe is just God under another name. But the divine name rejects this confusion. God is "I AM." The universe is not "I AM." The universe **exists**, but it does not exist by its own power. It receives existence. God alone is self-existent. Creation exists by participation, not by identity. The universe is real—genuinely, truly real—but its reality is derivative. God alone is necessary being. Everything else is contingent.[7]

This distinction preserves both the transcendence and the immanence of God. God is absolutely transcendent, utterly beyond creation, not limited by space or time or any creaturely category. Yet God is also immanently present, sustaining every moment of existence, upholding the structure of every atom, maintaining the order of every natural law. He is not distant and uninvolved, as deism imagines. He is the ground of being in whom all things live and move and have their being, as Paul declares to the Athenians (Acts 17:28, ESV). God remains transcendent, but creation reflects His rational structure. The universe is not God, but it is God's work, and it bears the marks of its Maker.

The divine name also illuminates human existence. You exist, but not independently. Your existence is received, moment by moment, from the One who is I AM. You are a contingent being whose identity is grounded in God's act of being. This dependence does not diminish your dignity. On the contrary, it establishes it. To be sustained by the One who is Being Itself is to possess a dignity that no earthly power can erase. Your worth is not rooted in your abilities, achievements, or circumstances. It is not based on your productivity, your intelligence, your health, or your social status. It is based on the fact that the eternal I AM wills you to exist, knows you by name, and sustains you in being at every instant. Your value is metaphysical. It is grounded in the nature of God.[8]

The personal dimension of the divine name is equally important. "I AM" is not a metaphysical abstraction. It is a personal self-identification. God speaks in the first person. He enters into relationship. He reveals Himself by name. In the ancient world, to know someone's name was to have access to them, to be able to call upon them, to enter into relationship with them. God gives Moses His name not to satisfy curiosity but to establish covenant. The God who is Being Itself becomes the God who is **with us**. The divine name is therefore both ontological and relational. God is the ground of existence **and** the source of communion.

This is why Jesus' use of the divine name is so significant. When He declares, "Before Abraham was, I AM" (John 8:58, ESV), He is not merely claiming temporal priority. He is identifying Himself with the eternal I AM who spoke to Moses from the burning bush. The religious authorities understand immediately what He is saying, that is why they pick up stones to stone Him for blasphemy. When Jesus says, "I AM the bread of life" (John 6:35), "I AM the light of the world" (John 8:12), "I AM the resurrection and the life" (John 11:25), "I AM the way, and the truth, and the life" (John 14:6), He is not simply offering metaphors. He is declaring that the fullness of being, life, truth, and existence is found in Him. The Logos through whom all things were made is the same I AM who revealed Himself to Moses. The Word became flesh. Being Itself entered creation.[9]

The divine name also shapes how you understand time and eternity. God's statement "I AM" is present tense, but it transcends temporal categories. God does not exist "in" time as creatures do. God is not subject to before and after, past and future. All moments are present to Him in the eternal now of His being. This does not mean God is frozen or timeless in the sense of being disconnected from temporal events. Rather, God's eternity includes and encompasses all of time without being limited by it. He interacts with creation in time. He speaks to Moses at a particular moment. He delivers Israel at a particular moment. He becomes incarnate in Christ at a particular moment, but His being is not stretched out through time like ours is. He simply is. And because He is I AM, He is the same yesterday, today, and forever (Hebrews 13:8, ESV).

Finally, the divine name invites you into a response. When God reveals Himself as I AM, He is not offering a philosophical proposition for academic consideration. He is calling Moses, and through Moses he is calling all of us into relationship with the living God. The proper response to "I AM" is not merely intellectual assent but trust, obedience, worship, and love. To know that God is I AM. Is to know that He is the ground of all being, the source of all existence, the One who holds you in being at every moment. It is to know that you are not your own. You are His. You exist because He wills it. You continue to exist because He sustains you. And He has revealed Himself not to remain distant but to draw you into communion with Him.

This chapter has shown that the divine name "I AM WHO I AM" is a statement of ontology, identity, and presence. It reveals God as Being Itself, the necessary ground of all contingent existence, the source of order and intelligibility in the universe, and the personal God who enters into covenant relationship with His creatures. The divine name connects the God of Scripture with the God of classical metaphysics, showing that these are not two different deities but one and the same. The God who spoke to Moses from the burning bush is the God through whom all things were made and in whom all things hold together. The next chapter explores how this understanding of God as I AM illuminates the Trinitarian nature of divine being and the relational structure of ultimate reality.

CHAPTER REVIEW:
KEY CONCEPTS:
- **Exodus 3:14 reveals God's name as Being Itself** Hebrew 'ehyeh 'asher 'ehyeh = "I AM WHO I AM" or "I WILL BE WHAT I WILL BE"

- **God's essence IS existence** Unlike creatures whose essence is distinct from existence, God is pure act of being
- **Self-sufficient and necessary** God depends on nothing; everything else depends on God
- **Personal and dynamic** "I AM" is not static abstraction but living, acting presence
- **Relational from eternity** God's being is inherently relational (Trinity), not isolation

WHAT THIS MEANS:

God's self-revelation as "I AM" is not just a name but a metaphysical disclosure. God is not one being among many but Being Itself, the source and ground of all existence. This connects classical metaphysics (pure act, necessary being) with biblical revelation, showing that ultimate reality is both metaphysically foundational and personally relational.

IMPORTANT TERMS & CONCEPTS:

- **I AM / YHWH** → Part X, Appendix B (Hebrew word studies)
- **Aseity (self-existence)** → Part X, Section 1.6 (metaphysical terms)
- **Divine Simplicity** → Part X, Section 1.6 (Divine Simplicity)
- **Pure Act (Actus Purus)** → Part X, Section 1.6 (Pure Act)
- **Necessary Being** → Part X, Section 1.6 (Necessary Being)

LOOKING AHEAD:

- Chapter 17 develops Trinitarian theology showing relationality in God's nature
- Chapter 18 explores implications for souls and afterlife
- Q&A Section I addresses questions about God's nature
- This metaphysical foundation supports all subsequent theological claims

TECHNICAL REFERENCE:

- For complete metaphysical terminology: Part X, Section 1.6 (all metaphysical terms)
- For divine attributes: Q&A Section I (Questions 1-10)
- For name of God theology: Appendix B (Hebrew studies); Chapter 16 endnotes

"I AM That I AM": God's Name and Ontology

- For classical theism concepts: Part X, Section 1.1 (Theology domain)

1. For the Hebrew text and translation options of **Ehyeh Asher Ehyeh**: Propp, W. H. C. (1999). *Exodus 1–18*. Anchor Bible Commentary. Doubleday, pp. 201–204. Durham, J. I. (1987). *Exodus*. Word Biblical Commentary. Word Books, pp. 37–39. The verb **ehyeh** is the imperfect first-person singular of *hayah*. The imperfect can indicate present, future, or continuous action. For the Septuagint translation **ego eimi ho on**: Wevers, J. W. (1990). *Notes on the Greek Text of Exodus*. Scholars Press, pp. 30–31. The LXX influenced early Christian theology profoundly, particularly in discussions of divine being.
2. For the metaphysical interpretation of Exodus 3:14 in Christian tradition: Gilson, É. (1941). *God and Philosophy*. Yale University Press, Chapter 2. Gilson traces how the Church Fathers and medieval scholastics understood the divine name as revealing God's essence as existence itself. Aquinas, T. *Summa Theologica*, Part I, Question 13, Article 11. For patristic exegesis: Gregory of Nyssa, *Against Eunomius* II.9–10; Augustine, *On the Trinity*, Books V–VII.
3. For ancient Near Eastern context and the uniqueness of YHWH: Tigay, J. H. (1986). *You Shall Have No Other Gods: Israelite Religion in the Light of Hebrew Inscriptions*. Harvard Semitic Studies 31. Scholars Press. Smith, M. S. (2001). *The Origins of Biblical Monotheism: Israel's Polytheistic Background and the Ugaritic Texts*. Oxford University Press. The revelation at the burning bush marks a decisive break from ANE polytheism by identifying God with being itself rather than with natural forces or localized powers.
4. For the doctrines of divine simplicity and aseity: Aquinas, T. *Summa Theologica*, Part I, Questions 3–4. Dolezal, J. E. (2011). *God Without Parts: Divine Simplicity and the Metaphysics of God's Absoluteness*. Pickwick Publications. Davies, B. (2004). *An Introduction to the Philosophy of Religion*, 3rd ed. Oxford University Press, Chapter 2. Divine simplicity has been debated in contemporary philosophy of religion, but it remains central to classical theism.
5. For the dynamic, covenantal nature of the divine name: Childs, B. S. (1974). *The Book of Exodus: A Critical, Theological Commentary*. Westminster Press, pp. 60–70. Fretheim, T. E. (1991). *Exodus*. Interpretation Commentary. John Knox Press, pp. 62–66. The burning bush theophany emphasizes both God's holiness (Moses must remove his sandals) and God's involvement in history (He hears the cry of His people and acts to deliver them).
6. For the cosmological argument from contingency: Leibniz, G. W. (1714). "The Principles of Nature and Grace, Based on Reason." In *Philosophical Essays*. Hackett, 1989, pp. 206–213. Feser, E. (2017). *Five Proofs of the Existence of God*. Ignatius Press, Chapter 2 (on the Leibnizian cosmological argument). The universe's contingency points to a necessary being whose essence includes existence.
7. For the distinction between God and creation, avoiding both materialism and pantheism: Gilson, É. (1956). *The Christian Philosophy of St. Thomas Aquinas*. University of Notre Dame Press, Part I, Chapter 3. Craig, W. L., & Moreland, J. P.

(2003). *Philosophical Foundations for a Christian Worldview*. InterVarsity Press, Chapters 26–27. Classical theism maintains both God's transcendence and His immanence without collapsing into pantheism or deism.

8. For human dignity grounded in divine creation and sustenance: Wolterstorff, N. (2008). *Justice: Rights and Wrongs*. Princeton University Press, Chapters 11–13. O'Donovan, O. (1994). *Resurrection and Moral Order*, 2nd ed. Eerdmans, pp. 13–30. Human worth is ontological, not functional—it derives from being sustained in existence by God, not from human capacities or achievements.

9. For Jesus' use of "I AM" statements in John's Gospel: Brown, R. E. (1966). *The Gospel According to John (I–XII)*. Anchor Bible Commentary. Doubleday, pp. 533–538 (on John 8:58). Bauckham, R. (2008). *Jesus and the God of Israel: God Crucified and Other Studies on the New Testament's Christology of Divine Identity*. Eerdmans, Chapter 6. Jesus identifies Himself with YHWH, the I AM of Exodus 3:14, claiming divine identity.

CHAPTER SEVENTEEN
FATHER, SON, SPIRIT: RELATIONALITY AS ULTIMATE REALITY

When you think about God, what comes to mind? Perhaps a distant creator who set the universe in motion and stepped back. Perhaps an all-powerful force governing the cosmos from afar. Perhaps a cosmic lawgiver who established the rules and now watches to see if we follow them. These images, while capturing something of divine transcendence and power, miss the heart of the Christian revelation. The God who reveals Himself as "I AM", the ground of all being, exists eternally not as solitary monarch but as communion: Father, Son, and Holy Spirit. The doctrine of the Trinity is not theological decoration added to make Christianity more mysterious. It is the key to understanding why the universe is relational, intelligible, ordered, and capable of producing persons who can know and love.

The deepest truth about reality is not isolated power but perfect communion. This changes everything about how you understand the world you inhabit and the life you live within it. The universe is relational because its Creator is relational. You are made for relationship because the God in whose image you are made exists eternally as relationship. Love is not a pleasant addition to existence, it is existence expressing its deepest nature. This chapter explores how the Trinity illuminates the structure of reality and reveals the meaning of personhood, consciousness, and the moral order.

Before going further, we must address what might seem like an obvious objection: isn't the Trinity a contradiction? How can God be both one and three? The confusion dissolves when we recognize that God is one in essence and three in person. Essence refers to **what** God is. Person refers to **who** God is. These are not competing claims but complementary truths about different aspects of divine reality.[1]

God's essence is the single, infinite act of being. There is one God, not three gods. This monotheism is fundamental to biblical faith. "Hear, O Israel: The LORD our God, the LORD is one" (Deuteronomy 6:4, ESV). The unity of God's being is absolute. But within this one divine essence exist three persons: Father, Son, and Spirit. These persons are not parts or fragments of God. Each person is fully God, sharing completely in the one divine essence. The Father is God. The Son is God. The Spirit is God. Yet the Father is not the Son, the Son is not the Spirit, and the Spirit is not the Father.

Think of it this way: the three persons are eternal relations within the one being of God. The Father knows and loves the Son. The Son knows and loves the Father. The Spirit is the bond of this eternal love, proceeding from the Father and resting upon the Son. This communion is perfect, infinite, and eternal. It is the fullness of being itself. The Trinity does not contradict unity, it shows that the highest unity is relational rather than solitary.[2]

This understanding of God has profound implications for everything you believe about reality. If the ground of being is relational, then relationality is woven into the fabric of creation. The universe does not arise from isolation but from communion. It is not built from self-contained, independent units but from relations. Modern physics confirms this insight in ways the ancient theologians could not have imagined. Quantum entanglement reveals that particles have their identity through relationships, not in isolation. Spacetime itself emerges from networks of correlation. Information theory shows that meaning arises from distinctions between states, distinctions that only make sense in relation to one another. The universe is relational at its foundations because its Creator is relational in His very being.[3]

The Trinity also illuminates what it means to be a person. You might think personhood is defined primarily by consciousness, by the capacity to think and feel. But Christian theology reveals something deeper: a person is fundamentally relational. To be a person is to exist in relation; to know, to love, to give, to receive. Isolated consciousness would not constitute personhood in its fullness. Persons become themselves through communion with others. This is not merely philosophical speculation. It reflects the nature of the God who made you. The Father is eternally Father because He eternally begets the Son. The Son is eternally Son because He is eternally begotten of the Father. The Spirit eternally proceeds from the Father and rests upon the Son. Each divine person is who He is through relationship with the others. And you, made in the image of this Triune God, are likewise constituted by your relationships. Your deepest experiences, like friendship, family, community, covenant, marriage, and worship are relational because they mirror the eternal relationality of your Creator.[4]

This means you were never meant to exist in isolation. The modern myth of the autonomous, self-made individual contradicts the deepest structure of reality. You discover who you are not by withdrawing from others but by entering into genuine communion with them. Your identity is shaped by those you love, those who love you, and ultimately by your relationship with the God who called you into being. The Trinity reveals that being itself is communion, and you become most fully yourself when you live in loving relationship with God and neighbor.

The Trinity also helps you understand one of the great mysteries of existence: how unity and diversity can coexist without contradiction. The Father, Son, and Spirit are perfectly one in essence, yet genuinely distinct in person. There is no confusion, no merging of identities, no loss of distinction. Yet there is also no division, no separation, no competing wills. Perfect unity contains real distinction. Perfect distinction preserves absolute unity. This harmony provides the metaphysical foundation for understanding diversity within unity throughout creation. Consider the universe itself: it contains countless forms of life, energy, and structure, yet all are unified by underlying laws and relations. Galaxies differ from one another, stars have individual characteristics, planets possess unique properties, living organisms exhibit stunning variety—and yet all of this diversity exists within a coherent, ordered whole. The universe mirrors the relational unity of its Triune Creator: many expressions, one reality.[5]

The same pattern appears in human community. You can see it in the church, which Paul describes as one body with many members. Each member has distinct gifts and callings, yet all belong to the one body of Christ. You can see it in a healthy society, where individuals maintain their particular identities while contributing to the common good. You can see it in marriage, where two persons become one flesh without losing their distinct personhood. Unity without diversity becomes tyranny or mere uniformity. Diversity without unity becomes fragmentation or chaos. The Trinity shows how these can be held together: genuine distinction within perfect communion.

Why does creation exist at all? The Trinity reveals that God creates not out of loneliness but out of generosity. The Father, Son, and Spirit exist in perfect, infinite, eternal communion. They lack nothing. They need nothing. Creation adds nothing to God's completeness or happiness. So why create? Because the fullness of divine life naturally overflows into the gift of existence to others. Love is inherently generative. It gives. It shares. It extends goodness beyond itself not from necessity but from abundance. The relational life of the Trinity extends outward in the act of creation. The universe is not a monument to divine power or an experiment in cosmic engineering. It is an expression of divine love, an invitation to finite beings to participate in the infinite communion of Father, Son, and Spirit.[6]

This means everything you see. Every star, every living thing, every moment of existence, is sustained by the Triune God. The Father is the source of all being. The Logos reveals the Father's wisdom and rationality in creation's order. The Spirit gives life, breath, and sanctification. Creation is not a machine set in motion and left to run. It is held in being at every instant by the active presence of the God who is Trinity. You exist within this divine activity. Every breath you take, every thought you think, every moment of consciousness occurs within the sustaining love of Father, Son, and Spirit.

The relational nature of the Trinity also explains why the universe produces beings capable of consciousness. If reality is grounded in relational being, in a God who knows and loves, then it makes perfect sense that the universe would eventually blossom into creatures capable of knowing and being known. Consciousness is not a cosmic accident, an inexplicable emergence from mindless matter. It is the natural flowering of a universe created by a God who is knowledge and love.[7]

Think about what consciousness requires: rationality, intentionality, the capacity for understanding and relationship. Where do these capacities come from? Materialist philosophy struggles to explain them, often resorting to the claim that consciousness simply "emerges" from complex arrangements of non-conscious parts. But emergence explains structure, not awareness. It describes new patterns of organization, not the qualitative experience of seeing red, feeling joy, or understanding truth. The Trinity offers a coherent explanation. Human rationality reflects the Logos, the divine Reason through whom all things were made. Human relationality reflects the Spirit, the bond of divine love. Human personhood reflects the Father's creative intention to make beings capable of entering into relationship with Him. You are not an accident of blind forces, but the intended fruit of a relational universe created by a relational God. Your capacity for thought, for love, for moral awareness, for transcendence; all of this makes sense when you understand that you were made in the image of the Triune God.

The Trinity also reveals the true nature of love. In contemporary culture, love is often reduced to emotion or sentiment, a feeling you have toward someone you find attractive or pleasant. But Christian theology shows that love is far more than feeling. It is the deepest truth of existence, the fundamental reality from which everything else derives. Love, as revealed in the Trinity, is the willing of good for another person. The Father eternally wills the good of the Son. The Son eternally wills the good of the Father. The Spirit is the active willing of their mutual good. This love is not needy or grasping. It is complete, abundant, self-giving. It is love that creates space for the other, that honors the other's distinctness, that never forces, manipulates, or controls.

This is the pattern of genuine love in human life. Love unites without destroying difference. It elevates without absorbing. It transforms without coercion. You see this in the best marriages, the deepest friendships, the most life-giving communities. Love allows the beloved to flourish as a distinct person while drawing them into deeper communion. This is possible because love reflects the nature of the Triune God, who is perfect unity without loss of distinction, perfect distinction without division.[8]

Love is also why you exist. You were not created as a utility or a tool but as a person capable of entering into relationship with the God who is love. Your highest calling is not achievement or success but participation in the divine communion of Father, Son, and Spirit. This means the most important thing about your life is not what you accomplish but who you become and whom you love. It means your worth is not measured by productivity or status but by your nature as one made in God's image and invited into His eternal life.

The relational nature of God also illuminates the moral order. Right action reflects participation in the divine relationality. Justice, mercy, forgiveness, and compassion are not arbitrary rules imposed from outside. They are expressions of the way reality is fundamentally structured. They flow from the nature of the Triune God. Consider justice: it honors the dignity and worth of persons, treating each as an end in themselves rather than as mere means. This reflects the Trinity, where each person is fully honored and loved for who they are. Consider mercy: it extends grace to those who have failed or fallen short, creating space for restoration and renewal. This reflects the divine generosity that continually sustains all creation in being despite its rebellion and brokenness. Consider forgiveness: it refuses to let wrongdoing define relationship permanently, offering instead the possibility of reconciliation and healing. This reflects the Father's heart toward wayward children, the Son's sacrifice for sinners, the Spirit's work of restoration.[9]

To act against love. To harm, to exploit, to dehumanize, or to dominate, is to act against the grain of being itself. It is to work against the fundamental structure of a relational universe created by a relational God. This is why such actions damage not only others but also yourself. You cannot flourish while violating the nature of reality. Conversely, to act in love. To serve, to honor, to lift up, and to bless; is to align yourself with the deep structure of existence. It is to participate in the life of the Triune God. This is not moralism or legalism. It is the recognition that virtue and vice are not arbitrary categories but reflect your compatibility or incompatibility with ultimate reality.

The Trinity also transforms your understanding of worship. Worship is not flattery designed to boost God's ego. God lacks nothing and needs no praise from creatures. Worship is not obligation; a duty you perform to avoid punishment. Worship is communion. It is participation in the divine life. When you pray, you are not shouting into the void hoping someone might hear. You are entering into the relational movement of Father, Son, and Spirit. The Spirit prays within you with groanings too deep for words, presenting your needs and longings to the Father through the mediation of the Son. When you love your neighbor, you are participating in the self-giving love that flows eternally within the Trinity. When you contemplate truth, beauty, or goodness, you are touching the reality of the Logos who orders all things. When you serve others sacrificially, you are reflecting the Son who "came not to be served but to serve, and to give his life as a ransom for many" (Matthew 20:28, ESV).[10]

This means worship is not confined to Sunday mornings or prayer closets. Every moment lived in love becomes an act of worship. Every kindness shown reflects the Trinity. Every truth discovered honors the Logos. Every beauty appreciated gives glory to the Creator. Worship becomes the highest expression of human identity. The conscious, willing participation in the communion of Father, Son, and Spirit that sustains all existence.

This chapter has shown that the Trinity is not an isolated doctrine, a peculiar Christian teaching disconnected from the rest of reality. It is the heart of Christian theology and the key to understanding the universe itself. God is Being Itself, and Being Itself is relational. The universe reflects this relationality in its fundamental structure. Persons embody it in their capacity for knowledge and love. Moral order expresses it in the call to justice and mercy. Worship realizes it in conscious communion with the divine life. You are not an isolated atom in a mechanistic universe. You are a person made in the image of the Triune God, invited into the eternal communion of Father, Son, and Spirit. You were created for relationship, for knowledge, for love. Your deepest longings for meaning, for connection, for transcendence; they are not random misfirings of neural circuits. They are the echo of the divine relationality from which you came and toward which you are called.

The next chapter will build on this foundation by exploring the nature of the soul, the spirit, and life beyond death. If human beings are made in the image of a relational God who exists beyond time and space, what does this mean for human destiny? What happens to consciousness when the body dies? How does the relational structure of the universe point toward the possibility, and indeed the promise, of continued existence in communion with the God who is life itself?

CHAPTER REVIEW:
KEY CONCEPTS:

- **Trinity: One essence (ousia), three persons (hypostases)** Not three gods nor one person in three modes
- **Being Itself is relational** God is not isolated monad but eternal communion
- **Persons defined by relations** Father, Son, Spirit exist in relation to each other eternally
- **Creation reflects Trinitarian pattern** Universe is relational because Creator is relational
- **Love is the nature of God** Not just an attribute but God's very essence (1 John 4:8)

WHAT THIS MEANS:

The Trinity reveals that ultimate reality is not abstract Being but relational communion. This has profound implications: persons (not just matter) are fundamental to reality; relationality (not isolation) is basic to existence; love (not power) is the deepest truth. The relational structure of quantum mechanics and the universe reflects its Trinitarian source.

IMPORTANT TERMS & CONCEPTS:

- **Trinity** → Part X, Q&A Section I; Section 1.1 (Theology)
- **Ousia (essence) vs. Hypostasis (person)** → Chapter 17 endnotes; Appendix B
- **Perichoresis (mutual indwelling)** → Chapter 17 endnotes
- **Person (theological definition)** → Part X, Section 1.1 (Theology)
- **Relationality** → Part X, Section 7.7 (Structural Resonance)

LOOKING AHEAD:

- Chapter 18 applies this to human souls and destiny
- Part VI connects Trinitarian theology to human dignity and freedom
- Q&A Sections develop implications for prayer, worship, salvation
- This is the theological capstone of the book's central argument

TECHNICAL REFERENCE:

- For Trinity doctrine: Q&A Section I; Part X, Section 1.1 (Theology)

- For person vs. nature: Part X theological concepts
- For relationality as structural principle: Part X, Section 7.7
- For common Trinitarian misunderstandings: Part X, Section 4

---

1. For the classical formulation distinguishing essence and person in Trinitarian theology, see the Athanasian Creed (5th–6th century) and the decrees of the Council of Constantinople (381). Contemporary exposition: Ware, K. (1997). *The Orthodox Way*, revised edition. St Vladimir's Seminary Press, Chapter 2. Ayres, L. (2004). *Nicaea and Its Legacy: An Approach to Fourth-Century Trinitarian Theology*. Oxford University Press, especially Chapter 11. The distinction between **ousia** (essence) and **hypostasis** (person) was critical in avoiding both modalism and tritheism while affirming both unity and distinction in God.
2. For the ontological relationality of the Trinity and its implications for understanding being itself, see: Zizioulas, J. D. (1985). *Being as Communion: Studies in Personhood and the Church*. St Vladimir's Seminary Press, especially Part I. LaCugna, C. M. (1991). *God For Us: The Trinity and Christian Life*. HarperSanFrancisco, Chapters 8–9. These works develop the patristic insight that personhood is constituted by relation, not by individual substance, and that this relational ontology derives from the Trinitarian nature of God.
3. For the connection between Trinitarian relationality and quantum entanglement as revealing the relational structure of physical reality, see: Polkinghorne, J. (2005). *Exploring Reality: The Intertwining of Science and Religion*. Yale University Press, Chapter 4. Russell, R. J., et al., eds. (2002). *Quantum Mechanics: Scientific Perspectives on Divine Action*. Vatican Observatory Publications, especially the essays by Polkinghorne and Peacocke. While quantum mechanics does not prove Trinitarian theology, the discovery that particles are defined by their relationships rather than possessing intrinsic, isolated properties resonates with the theological claim that being itself is relational.
4. [4] For the theological anthropology deriving from the Trinitarian nature of God, showing that persons are fundamentally relational beings made in the image of the relational God, see: Zizioulas, J. D. (1985). *Being as Communion*. Part II. Volf, M. (1998). *After Our Likeness: The Church as the Image of the Trinity*. Eerdmans, Chapter 4. McFadyen, A. I. (1990). *The Call to Personhood: A Christian Theory of the Individual in Social Relationships*. Cambridge University Press. Human personhood is not primarily defined by consciousness or rationality in isolation but by the capacity for relationship, reflecting the eternal communion of Father, Son, and Spirit.
5. For unity and diversity in creation reflecting Trinitarian unity and distinction, see: Gunton, C. E. (1993). *The One, the Three and the Many: God, Creation and the Culture of Modernity*. Cambridge University Press, especially Chapters 6–7. Torrance, T. F. (1996). *The Christian Doctrine of God: One Being Three Persons*. T&T Clark, Chapter 6. The perichoretic unity of the Trinity (mutual indwelling without confusion) provides the metaphysical foundation for understanding how the universe can contain genuine diversity within fundamental unity.

6. For creation as the overflow of Trinitarian love rather than divine necessity, see: Augustine, *The Trinity* (De Trinitate), Book V, Chapters 16–17, and Book XV. Contemporary treatment: Weinandy, T. G. (1995). *Does God Change? The Word's Becoming in the Incarnation* St. Bede's Publications, Chapter 3. Hart, D. B. (2003). *The Beauty of the Infinite: The Aesthetics of Christian Truth*. Eerdmans, pp. 155–249. God creates from abundance and generosity, not from need or compulsion, extending the gift of existence as an act of love.

7. For consciousness as the natural result of a universe created by a God who is knowledge and love, rather than an inexplicable emergence from non-conscious matter, see: Swinburne, R. (1997). *The Evolution of the Soul*, revised edition. Oxford University Press, Chapters 9–10. Taliaferro, C. (1994). *Consciousness and the Mind of God*. Cambridge University Press. Moreland, J. P., & Rae, S. B. (2000). *Body & Soul: Human Nature & the Crisis in Ethics*. InterVarsity Press, Chapters 3–4. These works argue that conscious, rational beings are more naturally explained by theism than by naturalism, particularly when the Creator is understood as Trinitarian—eternally knowing and loving.

8. For the nature of love revealed in the Trinity as the pattern for human love and relationship, see: Williams, R. (2000). *On Christian Theology*. Blackwell, Chapter 12 ("The Body's Grace"). Volf, M. (1996). *Exclusion and Embrace: A Theological Exploration of Identity, Otherness, and Reconciliation*. Abingdon Press, Part III. Pope Benedict XVI (2005). *Deus Caritas Est* (God is Love), Encyclical Letter, Part I. Love as revealed in the Trinity is self-giving, honors the distinct identity of the beloved, and creates communion without coercion or absorption.

9. For the grounding of moral order in the relational nature of the Triune God, see: O'Donovan, O. (1994). *Resurrection and Moral Order*, 2nd edition. Eerdmans, Chapters 1–2. Wolterstorff, N. (2008). *Justice: Rights and Wrongs*. Princeton University Press, Part IV. Adams, R. M. (1999). *Finite and Infinite Goods: A Framework for Ethics*. Oxford University Press, especially Chapters 1–2. Moral principles reflect the structure of reality as created and sustained by the God who is love, justice, and mercy in His very being.

10. For worship as participation in the divine life rather than obligation or flattery, and the Spirit's role in enabling human participation in Trinitarian communion, see: Torrance, J. B. (1996). *Worship, Community and the Triune God of Grace*. InterVarsity Press, especially Chapter 2. Schmemann, A. (1973). *For the Life of the World: Sacraments and Orthodoxy*, 2nd edition. St Vladimir's Seminary Press. Romans 8:26–27 (the Spirit interceding for us); Galatians 4:6 (the Spirit crying "Abba! Father!" in our hearts). Worship is our conscious participation through the Spirit in the Son's eternal love for the Father

CHAPTER EIGHTEEN
SOULS, SPIRITS, AND LIFE BEYOND DEATH

*I*f human beings are made in the image of a relational God, then you must be more than a physical organism. You possess an inward dimension that cannot be reduced to material processes. You are conscious, rational, moral, and capable of love. You experience yourself as an enduring subject who persists through time despite the constant turnover of your physical body. These features point toward the reality of the soul. This chapter examines the nature of the soul, the meaning of spirit, and the possibility of life beyond death in light of both classical theology and the relational universe described throughout this book.

The first step is to distinguish between soul and spirit. In Scripture, the soul refers to the life-principle that animates the body. It is the form of a living being, the organizing principle that gives unity to its parts. The spirit, by contrast, refers to the higher dimension of human life that opens the person to God. The soul makes a human being alive. The spirit makes a human being capable of communion with the divine. These distinctions are not rigid categories but aspects of a unified personal reality.[1]

From a philosophical standpoint, the soul is the principle that makes a human being one thing rather than a collection of parts. It is what allows the body to function as an integrated whole. Without the soul, the body disintegrates. The soul is therefore not a ghost inside the machine but the form that makes the body a living organism. This understanding aligns with the relational and ordered structure of the universe. Living beings maintain themselves by processing information, responding to their environments, and preserving their internal coherence. The soul is the principle that unites these processes into a single living subject.

The spirit goes further. You possess rationality, moral awareness, creativity, and the capacity to seek truth and goodness. These capacities cannot be reduced to biological instincts or physical processes. They point toward a dimension of human existence that transcends matter. Your spirit is oriented toward what is eternal, true, and good. It is capable of knowing God. This orientation reflects the divine image. The spirit is the aspect of your nature that corresponds to the relational life of Father, Son, and Spirit.[2]

The persistence of personal identity across time provides further evidence for the soul. Every cell in your body is replaced over the course of years, yet you remain the same person. Memory, intention, and self awareness continue. This continuity cannot be explained by physical structure, which constantly changes. It reflects the unity of the soul, which holds you together through time. You are not defined by matter but by form, meaning, and relational identity.

The existence of the soul also explains the human desire for transcendence. Across cultures and centuries, human beings long for meaning, purpose, and eternal life. This desire is not wishful fantasy. It is a reflection of the human spirit's orientation toward the infinite. As Augustine wrote, "Our hearts are restless until they rest in You."[3] Your yearning for the eternal mirrors the divine imprint within you. The soul is created for communion with God, and its deepest longing is to return to the Source of its being.

The question of life beyond death follows naturally. If the soul is the form of the body, what happens when the body dies? Classical theology teaches that the soul does not cease to exist. Because the soul is not material, it is not subject to physical dissolution. At death, the soul is separated from the body but continues to exist as the personal identity of the human being. This state is not natural in the fullest sense, because human nature is a unity of body and soul. The separation of the two is a temporary condition. The final destiny of the human person includes bodily resurrection, not disembodied existence alone.[4]

The possibility of life beyond death aligns with the structure of the universe described earlier in this book. Information in the physical world is never annihilated. It may transform or diffuse, but it is preserved. If the universe itself preserves informational identity, it is not unreasonable to believe that the Creator preserves the identity of persons. Moreover, the relational nature of reality suggests that death is not the final word. If the universe is grounded in a relational God, then relationships cannot be ultimately destroyed. Love, truth, and personal identity are not extinguished by death. They are held in the eternal knowledge of God.

The resurrection provides the Christian answer to the question of ultimate destiny. The resurrection of Christ is not merely a historical claim. It is a revelation of what human nature is meant to be. In the resurrection, the unity of body and soul is restored in a glorified form. The physical body becomes fully responsive to the spirit. The relational life of the Trinity becomes the destiny of redeemed humanity. The resurrection shows that human life is not a closed circle but an open journey toward communion with God.[5]

This understanding gives meaning to death itself. Death is not annihilation. It is transition. It is the point at which the soul returns to God for judgment, purification, and fulfillment. The Christian tradition affirms that each person stands before God with their freedom intact. The choices made in this life shape the soul's capacity for communion. The love, truth, and goodness you embrace become part of your soul's identity. The selfishness, falsehood, and sin you cling to must be purified. The soul enters eternity as the person it has become.

The existence of the soul also illuminates moral responsibility.

If persons were merely physical systems, moral choice would be an illusion. But because you possess a soul, your actions have lasting significance. Moral choices shape the soul. Acts of love, justice, and courage strengthen its capacity for communion. Acts of hatred, injustice, and cowardice wound it. The development of virtue is therefore the development of the soul's true form. Your vocation is to grow into the likeness of the One in whose image you are made.

The nature of the soul also reveals the purpose of creation. The universe is not merely the backdrop for physical processes. It is the environment in which souls come to maturity. The challenges of life are not meaningless obstacles. They are opportunities for growth, shaping the soul into a being capable of union with God. Suffering, while painful, can deepen compassion, humility, and love. Joy reveals the goodness of creation. Relationship teaches responsibility. Knowledge leads to wonder. Every aspect of life contributes to the formation of the soul.[6]

This chapter has shown that you are more than a material being. You are an embodied soul with a spirit oriented toward God. Your personal identity persists through time because it is grounded in the unity of the soul. Your destiny extends beyond death because your existence participates in the eternal act of God. The relational structure of the universe mirrors the relational nature of the soul. Your longing for eternity reflects the purpose for which you were created.

The next major section of the book turns outward, examining how the uniqueness of the Christian understanding of God, creation, and personhood distinguishes it from other religious and philosophical systems. If God is the eternal I AM and the universe reflects His relational nature, what makes Christianity uniquely suited to explain the structure of reality and the dignity of the human person?

CHAPTER REVIEW:
KEY CONCEPTS:

- **Soul as form/actuality of the living body** Aristotelian-Thomistic view; not Platonic dualism

- **Intermediate state between death and resurrection** Soul continues consciously; awaits bodily resurrection

- **Resurrection of the body** Not mere spiritual existence but transformed physical life

- **New heavens and new earth** Creation renewed, not discarded

- **Judgment as accountability** Real consequences for how we live; God's justice and mercy

WHAT THIS MEANS:

Christian hope is not escape from matter but resurrection of the body in a renewed creation. The soul's continuation after death is temporary; the final state is bodily. This honors both the spiritual dimension (souls continue) and the physical (matter matters to God). Accountability is real because persons are real and choices have meaning.

IMPORTANT TERMS & CONCEPTS:

- **Soul** → Part X, Q&A Section III; Section 1.5 (philosophical concepts)

- **Intermediate State** → Part X, Q&A Section VII (Questions 61-70)

- **Resurrection** → Q&A Section VII

- **New Creation** → Q&A Section IX (Questions 81-90)

- **Judgment** → Q&A Section IX

LOOKING AHEAD:

- Part VI develops how this grounds human dignity in THIS life

- Q&A Sections VII and IX address specific eschatological questions

- Chapter 22 explores living before I AM in light of eternal destiny

- This completes the theological anthropology begun in earlier chapters

TECHNICAL REFERENCE:

- For soul/body relationship: Q&A Section III (Questions 21-30)

- For afterlife questions: Q&A Section VII (Questions 61-70)

- For eschatology: Q&A Section IX (Questions 81-90)

- For what Christian hope IS and ISN'T: Part X, Section 3

1. For the classical distinction between soul and spirit in Christian anthropology, see: Aquinas, T. *Summa Theologica*. Part I, Question 75–76 (on the soul as substantial form); Question 77 (on the powers of the soul). Contemporary treatment: Cooper, J. W. (2000). *Body, Soul, and Life Everlasting: Biblical Anthropology and the Monism-Dualism Debate*, 2nd edition. Eerdmans. Moreland, J. P., & Rae, S. B. (2000). *Body & Soul: Human Nature & the Crisis in Ethics*. InterVarsity Press, Chapters 2–3. The soul (Hebrew **nephesh**, Greek **psyche**) is the animating principle of life; the spirit (Hebrew **ruach**, Greek **pneuma**) is the higher capacity for relationship with God.
2. For the transcendent capacities of the human spirit that cannot be reduced to material processes, see: Swinburne, R. (1997). *The Evolution of the Soul*, revised edition. Oxford University Press, Chapters 8–9. Taliaferro, C. (1994). *Consciousness and the Mind of God*. Cambridge University Press, Chapters 2–3. These works argue that rationality, moral awareness, and spiritual capacity point beyond purely physical explanation, suggesting the reality of an immaterial dimension of human existence.
3. Augustine of Hippo. *Confessions*, Book I, Chapter 1. Translation: Chadwick, H. (1991). *Augustine: Confessions*. Oxford University Press, p. 3. This famous opening captures the restlessness of the human soul seeking its rest in God, reflecting the innate orientation of human nature toward the transcendent and eternal.
4. For the classical Christian teaching on the intermediate state (the soul's existence between death and resurrection), see: Aquinas, T. *Summa Theologica*, Supplement, Questions 69–74. Contemporary discussion: Walls, J. L. (2002). *Heaven: The Logic of Eternal Joy*. Oxford University Press, Chapter 2. Wright, N. T. (2008). *Surprised by Hope: Rethinking Heaven, the Resurrection, and the Mission of the Church*. HarperOne, Chapters 2–5. The separated soul is conscious and retains personal identity but awaits the resurrection for the fullness of human nature to be restored.
5. For the resurrection of Christ as revealing the destiny of humanity and the transformation of the body, see: 1 Corinthians 15:35–58 (especially verses 42–49 on the resurrection body). Wright, N. T. (2003). *The Resurrection of the Son of God*. Fortress Press, Part VI. The resurrection is not mere resuscitation but the glorification and transformation of human nature into incorruptible, spiritual-physical form suited for eternal life in God's presence.
6. For the classical "soul-making" theodicy, viewing earthly life as the environment for moral and spiritual formation, see: Irenaeus of Lyons. *Against Heresies*, Book IV, Chapters 37–38. Modern development: Hick, J. (1966). *Evil and the God of Love*, revised edition (1977). Palgrave Macmillan, Chapters 13–17. While Hick's full universalist conclusions go beyond classical Christianity, the Irenaean emphasis on spiritual development through challenges remains a valuable framework for understanding the purpose of creation.

PART SIX
UNIQUENESS OF CHRISTIANITY AND THE MORAL UNIVERSE

CHAPTER NINETEEN
THE RELATIONAL GOD

The previous chapters have shown how modern physics, classical philosophy, and Christian theology converge on a coherent vision of reality. But a question remains: Why Christianity specifically? Many religions affirm the existence of God or ultimate reality. Many teach moral principles. Many inspire devotion and shape cultures. So, what makes Christianity unique in grounding the concepts of human dignity, individual rights, and ordered liberty that define Western civilization?

This is not a claim that Christianity is the only religion with valuable insights or that other traditions have nothing to offer. It is a claim about what specific theological commitments produce what specific political and moral consequences. Different understandings of God, humanity, and reality lead to different conclusions about how societies should be structured and how persons should be treated. Christianity's particular doctrines—the Trinity, the Incarnation, the image of God, natural law—have generated a distinctive vision of human dignity and freedom that other worldviews, however profound, do not produce in the same way.[1]

The foundation is the Christian doctrine of the image of God. Every human being, from conception to natural death, bears the image of God. This is not metaphorical. It is not limited to certain capacities or abilities. It is intrinsic to being human. The image of God means that you possess rationality, the capacity for relationship, moral agency, and creative power. These reflect the nature of God Himself. And because the image is inherent rather than earned, human dignity is universal and inalienable. It cannot be forfeited by disability, diminished by poverty, or revoked by the state.[2]

The Incarnation reinforces this beyond measure. In Jesus Christ, God became human. The eternal Logos took on flesh, lived as a man, experienced human limitations, suffered, died, and rose again in a glorified human body. This elevates humanity to an astonishing status. If God Himself entered creation by becoming human, then humanity is dignified in a way that transcends all other created beings. The body is not a prison. Matter is not evil. Physical existence is not something to escape but something God Himself embraced and will eternally retain in the resurrected Christ.[3]

This has direct implications for how you must treat other human beings. To harm a person is to attack the image of God. To deny someone's dignity is to despise what God has honored by taking it upon Himself. Every life—weak or strong, young or old, productive or dependent—possesses infinite worth because every life bears the divine image and belongs to the humanity that God assumed. This is the foundation for universal human rights, the prohibition against murder, the care for the vulnerable, and the principle that no person can be treated as mere means to another's ends.

The Trinity provides the foundation for relationality and personhood. God is not a solitary monad. God is three persons in perfect communion—Father, Son, and Holy Spirit—eternally giving and receiving love. This means that personhood is fundamentally relational. To be a person is not to exist in isolation but to exist in relationship. The Trinity models perfect unity in diversity, perfect communion without loss of distinction. And you, made in the image of this Triune God, are likewise made for relationship. You are not a self-sufficient atom. You are a person who finds fulfillment in community, in love, in mutual giving and receiving.[4]

This relational anthropology has profound political implications. It means that you have inherent worth, but you are not radically autonomous. You exist within networks of relationships—family, church, community, nation—that are natural and good. Liberty is not the absence of all constraint. It is the freedom to fulfill your nature within rightly ordered relationships. The state exists to protect persons and facilitate their flourishing, not to absorb them into a collective or to treat them as isolated individuals with no bonds or duties.

The doctrine of the fall explains why power must be limited. Christianity teaches that human nature is fallen. You are capable of great good, but also prone to selfishness, pride, and cruelty. This is not incidental. It is a fundamental truth about the human condition after the rupture of sin. The fall did not destroy the image of God, but it corrupted it. You retain rationality, relationality, and moral agency, but these capacities are distorted by sin. Left unchecked, human pride leads to tyranny. The concentration of power in fallible hands invites abuse.[5]

This is why the American Founders built a system of checks and balances. They did not assume that leaders would be virtuous or that power would be used justly. They assumed the opposite. They designed a government structure that pits ambition against ambition, dividing power to prevent any one person or faction from becoming tyrannical. This reflects a Christian understanding of human nature. We are capable of nobility, but we are also capable of great evil. Institutions must account for both. Power must be dispersed, limited, and held accountable.[6]

The concept of conscience is also distinctly Christian. The New Testament teaches that God's law is written on the heart, that every person has an inner sense of right and wrong that testifies to moral truth. Conscience is not infallible, it can be seared or misinformed, but it is real. And it is sacred. The state has no authority over the conscience. Faith must be voluntary. Belief cannot be coerced. This principle, articulated clearly by Christian thinkers and enshrined in the First Amendment, is the foundation of religious liberty. And once you recognize that the state cannot compel belief, you recognize that there are spheres of life—family, religion, conscience—that lie beyond the legitimate reach of government.[7]

Natural law theory, developed most fully by Christian theologians like Thomas Aquinas, provides the framework for objective morality. Natural law is the participation of rational creatures in the eternal law of God. It is accessible to reason by reflecting on human nature and the created order. Certain moral truths are self-evident to properly functioning reason: it is wrong to murder, to lie, to steal. These are not arbitrary rules. They are grounded in the nature of reality itself. And because natural law is objective, it provides a standard by which human laws can be judged. A law that contradicts natural law is not truly law. It is a perversion of law and does not bind the conscience.[8]

This gives you and your community a basis for resisting tyranny. If government enacts unjust laws, laws that violate human dignity or command evil, those laws are null and void. Obedience is not owed. This is not chaos. It is the recognition that human authority is limited, that rulers are subject to a higher law, and that when earthly authorities command what God forbids or forbid what God commands, allegiance belongs to God. This principle was invoked by the American Founders in the Declaration of Independence, by abolitionists resisting slavery, by the civil rights movement confronting segregation, and by Christians throughout history standing against unjust regimes.[9]

Christianity also affirms the goodness of creation and the legitimacy of temporal life. Unlike some Eastern religions that see the material world as illusion or some Gnostic heresies that view matter as evil, Christianity teaches that creation is good. God made the physical world and declared it good. The Incarnation confirms this. God entered the material order, lived in a body, and rose bodily from the dead. This means that work, family, politics, art, science: all of these belong to the good life. The spiritual and the material are not opposed. They are integrated. You are called to live faithfully in this world, to steward creation, to build just societies, to raise families, to create beauty. The Christian vision is not escapist. It is engaged, affirming that temporal life matters because God cares about His creation and is working to redeem it.[10]

Now consider how other worldviews differ, not to disparage them but to show that different theological commitments lead to different political conclusions. Islam affirms one God and shares with Christianity the belief in creation, moral law, and accountability. But Islamic theology does not develop the Trinity, and therefore it does not ground relationality in the divine nature the same way Christianity does. More significantly, traditional Islamic political thought does not separate religious and civil authority. Sharia is comprehensive law governing all aspects of life. While there is diversity within Islamic thought and many Muslims live peacefully in pluralistic societies, classical Islamic teaching does not provide the same foundation for limited government or freedom of conscience that Christianity does. The concept of rights preceding government and the idea of spheres of life beyond state control are not native to traditional Islamic jurisprudence.

Hinduism offers profound spiritual insights and a rich philosophical tradition. But classical Hindu metaphysics does not affirm the same kind of personal Creator or the same intrinsic dignity for every individual. The caste system, rooted in Hindu cosmology, assigns people to hierarchical social positions based on birth. While modern Hindu thinkers have challenged caste, and while Gandhi drew on both Hindu and Christian sources to oppose it, the caste system reflects a worldview in which human value is differentiated rather than universal. Buddhism teaches compassion and the alleviation of suffering, but it does not ground dignity in the nature of a personal God who creates humans in His image. Instead, the self is ultimately an illusion to be transcended. This makes it difficult to ground inalienable rights in a metaphysical reality that affirms the enduring value of the person.

Secular humanism attempts to preserve Christian moral commitments without Christian metaphysics. It affirms human dignity, equality, and rights but grounds them in human consensus or utility rather than in the nature of God. The problem is that without a transcendent foundation, these values become unstable. If humans are nothing more than complex physical systems produced by blind evolutionary processes, why should any individual possess inherent worth? Why should rights be inalienable? Why should power be constrained? Secular humanism borrows the moral capital of Christianity without acknowledging the source. And as Western societies drift further from their Christian roots, we see these borrowed values eroding. Rights are increasingly seen as social constructs. Truth is treated as relative. Human dignity becomes conditional. Moral claims are dismissed as preferences.[11]

The evidence of history supports this claim. Societies shaped by Christian theology developed concepts of human rights, constitutional limits on power, separation of church and state, care for the vulnerable, and the rule of law. These did not emerge from secular philosophy or from other religious traditions in the same way. The abolition of slavery, the protection of children, the establishment of hospitals and universities, the development of international humanitarian law: all of these bear the marks of Christian influence. This is not to say that Christians have always lived up to Christian ideals. They have not. Christians have participated in slavery, oppression, and injustice. But when they did so, they violated their own principles. The reforms came from within the Christian tradition, from people who appealed to Christian teachings to condemn Christian failures.[12]

This is why secular regimes, when untethered from Christian moral capital, have often devolved into horrors. The twentieth century's totalitarian movements were explicitly atheistic and treated humans as expendable means to ideological ends. Christianity provides what other systems lack: a personal God who is Being Itself, who creates you in His image, who reveals moral law, who enters creation to redeem it, and who establishes your worth as infinite and non-negotiable. This is not cultural imperialism. It is recognition of what ideas produce what consequences. The concepts that define the free world—liberty, equality, justice, human rights—are not universal constants that any culture would discover. They are the fruit of a specific theological tradition. They emerged from Christian soil, were nurtured by Christian thinkers, and were institutionalized in societies shaped by Christian convictions.[13]

This does not mean non-Christians cannot affirm these values. Of course they can. Many do. But the question is whether those values can be sustained without their original foundation. The evidence suggests they cannot. As Western societies abandon Christianity, they are also losing confidence in the concepts Christianity grounded. The result is not enlightened secularism. It is moral confusion, social fragmentation, and the rise of ideologies that deny the very principles the West once held dear.

If we want to preserve liberty, we must preserve the ideas that make liberty intelligible. If we want to protect human dignity, we must acknowledge its source. And if we want future generations to enjoy the freedoms we inherited, we must teach them why those freedoms exist. This is not a call to theocracy. It is a call to intellectual honesty. The American experiment rests on theological foundations. The principles we cherish are Christian in origin. And no secular substitute has proven capable of bearing the same weight.

The uniqueness of Christianity is not in its exclusivity but in its explanatory power. It provides answers to the deepest human questions—Why do I exist? What is my purpose? Why does my life matter?—that cohere with the structure of reality revealed by science and philosophy. It offers a vision of God that is both transcendent and personal, a vision of humanity that is both dignified and fallen, and a vision of society that balances individual liberty with communal responsibility. No other worldview integrates these dimensions as fully or as coherently.

The next chapter examines the moral law more deeply, showing how Christian anthropology provides the foundation for objective morality that applies across all domains—physical, philosophical, and theological—and why this grounding is essential for a just and flourishing society.

CHAPTER REVIEW:

KEY CONCEPTS:

- **Imago Dei as foundation for human rights** Each person bears God's image; dignity is intrinsic, not earned

- **Theological roots of Western liberty** From Aquinas → Salamanca School → Locke → American Founding

- **Natural law accessible to reason** Moral order discoverable through rational reflection, not just revelation

- **Secular substitutes lack grounding** Rights as "social construct" or "evolutionary byproduct" provide no stable foundation

- **Christianity's unique contributions** Universal human dignity, individual conscience, moral equality, limits on state power

WHAT THIS MEANS:

The values defining Western civilization—human rights, individual liberty, equal dignity—are not self-evident truths that any culture would discover. They emerged from specific Christian theological commitments about imago Dei and natural law. Removing the theological foundation while keeping the values is intellectually unstable; the tree cannot survive when roots are cut.

IMPORTANT TERMS & CONCEPTS:

- **Imago Dei** → Q&A Section III; Part X theological concepts

- **Natural Law** → Chapter 20; Part X, Section 1.1 (Philosophy domain)

- **Salamanca School** → Chapter 19 endnotes

- **Dignity** (intrinsic vs. attributed) → Q&A Section III

- **Rights** (unalienable vs. granted) → Chapter 19 discussion

LOOKING AHEAD:

- Chapter 20 develops moral law across physics, philosophy, theology

- Chapter 21 examines Christianity's role in building Western civilization

- This grounds the practical implications developed in Part VII
- Q&A sections address specific applications

TECHNICAL REFERENCE:

- For natural law tradition: Part X, Section 1.1 (Philosophy); Chapter 20 endnotes
- For imago Dei theology: Q&A Section III (Questions 21-30)
- For political philosophy connections: Chapter 19 endnotes (extensive citations)

---

1. For the argument that Christianity's specific theological commitments uniquely ground Western concepts of rights and liberty, see: Wolterstorff, N. (2008). *Justice: Rights and Wrongs*. Princeton University Press, especially Part III on the Christian foundations of rights. Siedentop, L. (2014). *Inventing the Individual: The Origins of Western Liberalism*. Belknap Press of Harvard University Press. Siedentop traces how Christian theological anthropology, particularly the concept of the soul, transformed Western understandings of personhood and political order.
2. For the doctrine of the **imago Dei** (image of God) as the foundation of universal human dignity, see: Genesis 1:26-27; Psalm 8:4-6; James 3:9. Theological treatment: Middleton, J. R. (2005). *The Liberating Image: The Imago Dei in Genesis 1*. Brazos Press. The image is inherent, not functional—it does not depend on rationality, productivity, or any other capacity, which is why it grounds the dignity even of those who lack such capacities.
3. For the Incarnation as affirming and elevating human nature, see: Athanasius of Alexandria. *On the Incarnation*, Sections 8-10 (4th century). Contemporary treatment: Torrance, T. F. (1981). *Divine and Contingent Order*. Oxford University Press, Chapter 4. Gunton, C. E. (1992). *Christ and Creation*. Eerdmans. By assuming human nature, the Logos permanently joins divinity and humanity, establishing the dignity of embodied existence and the goodness of matter.
4. For the Trinitarian foundation of relational personhood and its political implications, see: Zizioulas, J. D. (1985). *Being as Communion: Studies in Personhood and the Church*. St Vladimir's Seminary Press. Volf, M. (1998). *After Our Likeness: The Church as the Image of the Trinity*. Eerdmans, Chapters 4-5. The perichoretic communion of Father, Son, and Spirit provides the theological basis for understanding persons as inherently relational rather than atomistically autonomous.
5. For the doctrine of the fall and its implications for political realism, see: Genesis 3 (the fall narrative); Romans 3:23; 5:12-21 (sin's universal scope). Political application: Niebuhr, R. (1941). *The Nature and Destiny of Man*, Vol. 1. Charles Scribner's Sons, especially Chapters 7-9. Niebuhr argues that Christian realism about sin grounds a politics that limits power and checks human pride, avoiding both utopianism and cynicism.

6. For the American Founders' use of checks and balances reflecting Christian anthropology, see: Madison, J. (1788). *Federalist No. 51*. In *The Federalist Papers*. Madison's famous statement "If men were angels, no government would be necessary" reflects the Christian teaching that fallen human nature requires institutional constraints. See also: Dreisbach, D. L., & Hall, M. D., eds. (2014). *The Sacred Rights of Conscience: Selected Readings on Religious Liberty and Church-State Relations in the American Founding*. Liberty Fund.

7. For the Christian foundations of conscience and religious liberty, see: Romans 2:14–15 (law written on the heart); Acts 5:29 ("We must obey God rather than men"). Theological development: Aquinas, T. *Summa Theologica*, I-II, Questions 19 (on conscience) and 96 (on unjust laws). Modern application: Witte, J., Jr. (2000). *Religion and the American Constitutional Experiment*, 2nd edition. Westview Press. The concept of an inviolable inner forum of conscience, beyond state authority, is distinctly Christian.

8. For natural law theory in Christian thought, see: Aquinas, T. *Summa Theologica*, I-II, Questions 90–94 (Treatise on Law). Contemporary exposition: Finnis, J. (2011). *Natural Law and Natural Rights*, 2nd edition. Oxford University Press. George, R. P. (1999). *In Defense of Natural Law*. Oxford University Press. Natural law provides an objective moral standard grounded in human nature and accessible to reason, independent of revelation, yet ultimately rooted in God's eternal law.

9. For resistance to unjust authority based on higher law, see: Acts 5:29; Romans 13:1–7 (authority from God, but implicitly limited to just governance). Aquinas, *Summa Theologica*, I-II, Q. 96, Art. 4 (unjust laws do not bind in conscience). Historical applications: Declaration of Independence (1776) appeal to "Laws of Nature and of Nature's God"; King, M. L., Jr. (1963). "Letter from Birmingham Jail," invoking natural law and Aquinas to justify civil disobedience against segregation.

10. For the Christian affirmation of creation's goodness and the legitimacy of temporal life, see: Genesis 1:31 ("God saw everything that he had made, and behold, it was very good"); 1 Timothy 4:4 ("everything created by God is good"); Colossians 1:16 (all things created through Christ). Contrast with Gnosticism: Jonas, H. (1958). *The Gnostic Religion: The Message of the Alien God and the Beginnings of Christianity*, 2nd edition. Beacon Press. Christianity's affirmation of matter grounds engagement with the world rather than escape from it.

11. For the critique of secular humanism's attempt to retain Christian values without metaphysical foundation, see: Nietzsche, F. (1882/1974). *The Gay Science*, Section 125 ("God is dead" and the consequences). MacIntyre, A. (1981). *After Virtue: A Study in Moral Theory*. University of Notre Dame Press, Chapter 2. Secular liberalism borrows moral capital from Christianity but lacks the metaphysical resources to sustain it once the theological foundation erodes.

12. For the historical record of Christian influence on humanitarian reforms, see: Stark, R. (1996). *The Rise of Christianity: How the Obscure, Marginal Jesus Movement Became the Dominant Religious Force in the Western World in a Few Centuries*. Princeton University Press. Holland, T. (2019). *Dominion: How the Christian Revolution Remade the World*. Basic Books. Scrivener, G. (2022). *The Air We Breathe: How We All Came to Believe in Freedom, Kindness, Progress, and Equality*. The Good

Book Company. These works trace how Christian theology shaped Western moral and political culture.
13. For the argument that modern secular ideologies, untethered from Christian foundations, have led to totalitarianism, see: Solzhenitsyn, A. I. (1973). *The Gulag Archipelago*. Harper & Row (especially on atheistic Marxism's consequences). Arendt, H. (1951). *The Origins of Totalitarianism*. Harcourt Brace. While Arendt's analysis is complex, she notes the collapse of traditional moral constraints in totalitarian regimes, which often explicitly rejected Christian anthropology.

CHAPTER TWENTY
THE MORAL LAW ACROSS THREE DOMAINS

Morality is not invented. It is discovered. This claim stands in stark contrast to much of contemporary thought, which treats moral values as preferences, social constructs, or evolutionary adaptations with no objective truth. But if morality is merely subjective, then moral disputes are nothing more than conflicts of taste. There is no ground for saying that justice is better than injustice, that compassion is superior to cruelty, or that human dignity must be respected. Yet you do make such claims. You insist that some actions are truly wrong and others truly right, regardless of personal opinion or cultural context. The question is: What grounds this insistence?

Christian theology provides an answer that integrates insights from physics, philosophy, and revealed truth. Morality is grounded in the nature of God and the structure of reality He created. It is not arbitrary. It is not changeable. It flows from who God is and what it means to be human. And because the same God who reveals Himself in Scripture is the God who created the universe and endowed you with reason, moral truth is accessible through three complementary pathways: the natural order studied by science, the rational reflection of philosophy, and the explicit teaching of theology. These three domains do not contradict. They illuminate the same moral reality from different angles.[1]

Physics reveals a universe that is ordered, lawful, and intelligible. The regularities you observe—conservation of energy, symmetry principles, causal relationships—are not accidents. They reflect a deep structure built into reality. This structure is not itself moral in the sense of making value judgments, but it provides the stable framework within which moral agents can act. A universe governed by chaos would be uninhabitable. A universe without causation would make responsibility meaningless. The fact that actions have predictable consequences, that choices matter, and that the physical world responds consistently to your intentions is the precondition for moral life.[2]

The directionality of time also has moral implications. The arrow of time means that the past is fixed and the future is open. What has been done cannot be undone. Consequences unfold irreversibly. This creates the conditions for moral responsibility. If you could simply reverse time and undo your actions, moral choice would be meaningless. But because time moves forward and actions have lasting effects, your decisions carry weight. The universe's temporal structure makes morality possible and significant.

The relational structure of reality, revealed by quantum mechanics and information theory, also supports moral understanding. You exist within networks of relationships. Your actions affect others. Your choices create ripples that extend beyond yourself. This relational fabric is not morally neutral. It reflects the fact that you were made for communion, that you flourish in right relationships and suffer in broken ones. The universe is structured in a way that rewards cooperation, trust, and love, and punishes isolation, deception, and cruelty—not because of divine intervention in every moment, but because these patterns are woven into the nature of relational being.[3]

Philosophy deepens this understanding. Natural law theory, developed most fully in the Christian tradition by thinkers like

Thomas Aquinas, holds that certain moral principles are accessible to reason by reflecting on human nature and the created order. What does it mean to be human? What are our natural inclinations? What fulfills us? Reason can discern that humans naturally seek knowledge, form families, live in communities, and pursue the good. From these observations, philosophy derives moral principles: preserve life, educate the young, seek truth, live in justice, worship what is ultimate. These are not arbitrary rules imposed from outside. They are the conditions for human flourishing discovered through rational reflection on what we are.[4]

Natural law is objective because it is grounded in the nature of reality itself. It does not change with cultural preferences or individual desires. A society that honors these principles will flourish. A society that violates them will decay. History bears this out. Civilizations that protect the vulnerable, uphold justice, encourage virtue, and restrain vice tend to endure and prosper. Civilizations that exploit the weak, embrace injustice, reward vice, and punish virtue collapse into chaos. This is not coincidence. It is the natural consequence of aligning with or violating the moral structure of reality.[5]

Theology completes the picture. While physics reveals the conditions for moral action and philosophy identifies principles of natural law, theology reveals the source and ultimate ground of morality in the character of God. God is not merely powerful or knowledgeable. He is good, just, merciful, and loving. These are not arbitrary attributes. They are essential to who God is. And because God is the Creator, His character is reflected in the structure of creation. The moral law is not external to God, imposed on Him by some higher standard. It flows from His nature. God commands what is good because He is good. What aligns with God's nature is right. What contradicts it is wrong.[6]

This grounding in divine nature gives morality its binding force. You are not free to disregard moral law any more than you are free to disregard physical law. Just as jumping off a building will not end well regardless of your beliefs about gravity, so violating moral law will damage you regardless of your opinions about right and wrong. This is not arbitrary divine command. It is the recognition that reality itself is structured morally because it comes from a moral God.

Scripture reveals specific moral commandments that clarify and reinforce what reason can discover through natural law. The Ten Commandments, the teachings of Jesus, the ethical instructions of the apostles: all of these provide concrete guidance for how to live in accordance with God's will. But Scripture does not introduce morality for the first time. It reveals more fully and clearly what is already written on the human heart. This is why people from diverse cultures and times recognize murder, theft, and betrayal as wrong even without access to biblical revelation. The moral law is universal because it is grounded in the nature of God and reflected in the nature of humanity.[7]

One of the most profound aspects of Christian moral teaching is the integration of law and grace. The moral law reveals what is right, but it also exposes human failure. You know what you ought to do, but you do not always do it. You recognize justice, but you do not always act justly. This gap between knowledge and action is the reality of sin. Christianity does not deny this gap or minimize it. Instead, it offers grace. God does not lower the standard to accommodate failure. He provides the means to meet it. Through the life, death, and resurrection of Jesus Christ, forgiveness is offered and transformation is made possible. The Holy Spirit empowers you to live in accordance with the moral law, not through mere human effort but through participation in the divine life.[8]

This transforms morality from external obligation to internal transformation. You do not obey out of fear of punishment but out of gratitude for grace. You do not strive to earn God's favor but respond to the favor already given. Morality becomes the natural expression of a heart renewed by the Spirit. This is why Jesus summarized the entire law in two commandments: love God with all your heart, and love your neighbor as yourself. Love is not opposed to law. It is the fulfillment of law. When you truly love, you do not need a list of rules to tell you not to harm others. Love itself guides you toward what is good, true, and just.[9]

The moral law is also Christocentric. Jesus is not merely a teacher of morality. He is the embodiment of it. To see Jesus is to see what a fully moral life looks like. He is compassionate without being sentimental. He is just without being harsh. He is truthful without being cruel. He is strong without being domineering. He is humble without being weak. The virtues are unified in Him. And the goal of the Christian life is to be conformed to His image. Morality is not abstract principles. It is the imitation of Christ.[10]

The integration of physics, philosophy, and theology in grounding morality is powerful. Physics reveals a universe that is ordered and purposeful, providing the stable framework for moral action. Philosophy identifies the principles of natural law and the conditions for human flourishing, accessible to reason. Theology reveals the source of the moral law in the character of God, the ultimate ground of obligation, and the gracious provision for redemption when you fail. Together, these three domains show that morality is not arbitrary, not subjective, not a human invention. It is woven into the fabric of reality because reality comes from a moral God.

This has immense practical importance. If morality is objective, then moral progress is real. Societies can become more just. Individuals can become more virtuous. Evil can be resisted and overcome. But if morality is subjective, then there is no progress, only change. There is no justice, only power. There is no good or evil, only preferences. The stakes could not be higher. The choice is between a world where moral claims are meaningful and binding, or a world where might makes right and every person does what is right in their own eyes.

The West was built on the belief that morality is objective, that it is grounded in God, and that it applies to all people. This belief made possible the concepts of human rights, the rule of law, and the dignity of persons. It drove the abolition of slavery, the protection of the vulnerable, and the defense of liberty. As these theological foundations erode, so do the moral principles they support. The recovery of a robust Christian understanding of the moral law is not just a religious concern. It is a civilizational necessity. Without it, the ideals that have shaped our societies lose their coherence and their power to command assent. With it, you have a vision of morality that is intellectually defensible, practically livable, and ultimately grounded in the nature of ultimate reality itself.

The next chapter builds directly on this foundation by examining how Christianity's understanding of moral law shaped the institutions, values, and social practices that define Western civilization. If objective morality grounded in God's character is the key to human flourishing, how has this truth worked itself out in history? What happens when societies embrace it? And what happens when they abandon it?

CHAPTER REVIEW:
KEY CONCEPTS:

- **Physics reveals ordered, lawful universe** Stable structure allows moral action; physical regularities parallel moral order

- **Philosophy identifies natural law** Reason can discover moral principles accessible to all

- **Theology reveals God as source** Moral law grounded in divine character; ultimate obligation to personal God

- **Objective morality vs. subjectivism** Moral realism: some things are genuinely right/wrong independent of opinion

- **Integration across domains** Physics (framework), philosophy (reasoning), theology (ground)

WHAT THIS MEANS:

Morality is not invented but discovered because it reflects the structure of reality created by a moral God. All three domains contribute: physics shows the ordered cosmos that makes moral life possible; philosophy articulates moral principles through reason; theology reveals why moral obligations are binding. Objective morality provides the only coherent foundation for human rights and justice.

IMPORTANT TERMS & CONCEPTS:

- **Natural Law** → Part X, Section 1.1 (Philosophy); Chapter 20 endnotes
- **Moral Law** → Part X, Q&A Section applicable to morality
- **Objective Morality** → Part X, Section 1.1 (Philosophy domain)
- **Is-Ought Problem** → Chapter 5; Part X, Section 1.1
- **Divine Command Theory (and alternatives)** → Chapter 20 endnotes

LOOKING AHEAD:

- Chapter 21 shows how Christian moral vision shaped Western institutions
- Part VII applies these principles to personal life
- Q&A sections address specific moral questions
- This completes the three-domain framework applied to ethics

TECHNICAL REFERENCE:

- For natural law comprehensive treatment: Chapter 20 endnotes
- For moral philosophy: Part X, Section 1.1 (Philosophy domain)
- For integration of domains: Part X, Section 7 (Integration Guidelines)
- For is-ought gap: Part X, Section 1.1; Chapter 5

---

1. For the integration of physics, philosophy, and theology in grounding objective morality, see: Aquinas, T. *Summa Theologica*, I-II, Questions 90–94 (Treatise on Law), especially Q. 91 on the varieties of law (eternal, natural, human, divine). Modern synthesis: Feser, E. (2009). *Aquinas: A Beginner's Guide*. Oneworld Publications, Chapter 5. The natural law tradition holds that morality is grounded in the structure of reality, accessible to reason, and ultimately rooted in God's eternal law.

2. For the moral implications of physical order and causation, see: Polkinghorne, J. (1998). *Belief in God in an Age of Science*. Yale University Press, Chapter 3. The regularity and intelligibility of natural law make rational moral agency possible; a chaotic universe would undermine responsibility.

3. For the relational structure of reality and its moral implications, see: Zizioulas, J. D. (1985). *Being as Communion*. St Vladimir's Seminary Press. The claim is not that physics proves morality but that the relational character of the physical universe is congruent with a moral order grounded in a relational God. See also: Polkinghorne, J. (2005). *Exploring Reality*, Chapter 4.

The Moral Law Across Three Domains 219

4. For Thomistic natural law theory and its derivation from human nature, see: Aquinas, T. *Summa Theologica*, I-II, Q. 94, Arts. 2–4 (on the precepts of natural law). Contemporary restatement: Finnis, J. (2011). *Natural Law and Natural Rights*, 2nd edition. Oxford University Press, Chapters 3–4. Finnis identifies basic human goods (life, knowledge, play, aesthetic experience, sociability, practical reasonableness, religion) as self-evident starting points for moral reasoning.
5. For the historical correlation between adherence to natural law principles and societal flourishing, see: Stark, R. (2005). *The Victory of Reason: How Christianity Led to Freedom, Capitalism, and Western Success*. Random House. While causation is complex, Stark argues that Christian moral and legal principles contributed significantly to Western economic and political development.
6. For divine command theory grounded in God's nature rather than arbitrary will, see: Adams, R. M. (1999). *Finite and Infinite Goods: A Framework for Ethics*. Oxford University Press, Chapters 1–2, 13–14. Adams argues that moral goodness is grounded in resemblance to God, who is the Good itself, thus avoiding the Euthyphro dilemma.
7. For the universality of moral law and its expression in Scripture and conscience, see: Romans 2:14–15 (Gentiles who do not have the law do by nature what the law requires, showing the work of the law written on their hearts). Lewis, C. S. (1943). *The Abolition of Man*. Oxford University Press, Appendix (illustrations of the natural law from diverse cultures). Lewis demonstrates widespread moral consensus across civilizations, pointing to an objective moral order.
8. For the integration of law and grace in Christian ethics, see: Romans 3:20–26 (justification by faith apart from works of the law); Romans 8:1–4 (the law's requirements fulfilled in those who walk by the Spirit). Theological treatment: Barth, K. (1957). *Church Dogmatics* II/2: *The Doctrine of God*, T&T Clark, §36–37 (The Command of God).
9. For the summary of the law in love, see: Matthew 22:37–40; Romans 13:8–10; Galatians 5:14. Theological development: Augustine, *On the Spirit and the Letter* (412–415 AD). Contemporary treatment: O'Donovan, O. (1994). *Resurrection and Moral Order*, 2nd edition. Eerdmans, Chapters 11–13. Love is not opposed to law but is its fulfillment and inner principle.
10. For Christ as the embodiment of moral perfection and the pattern for Christian ethics, see: Philippians 2:5–11 (the mind of Christ); 1 Peter 2:21–23 (Christ as example); 1 John 2:6 (walking as Christ walked). Bonhoeffer, D. (1959). *The Cost of Discipleship*. Macmillan. Hauerwas, S. (1975). *Character and the Christian Life: A Study in Theological Ethics*. Trinity University Press. Christian ethics is fundamentally about Christoform existence—becoming conformed to the image of Christ.

CHAPTER TWENTY-ONE
THE HEART OF WESTERN CIVILIZATION

Many of the moral assumptions that shape modern Western life feel so natural that they rarely draw attention. That every human life has value. That power should justify itself. That the weak deserve protection. That rulers are accountable. That love is not merely sentiment but obligation. These ideas feel less like achievements than like background conditions, simply "how society works." But history tells a different story.

For most of human civilization, these assumptions were neither obvious nor widespread. They did not arise automatically from reason, biology, or social efficiency. They emerged slowly, unevenly, and often against the grain of prevailing cultural norms. And they emerged, again and again, in places shaped by the life and teaching of Jesus of Nazareth. This chapter is not an argument that Christian societies have always lived up to Christian ideals—they have not. Nor is it a claim of moral superiority. It is, instead, an attempt to trace how certain ideas that now feel universal entered the moral bloodstream of Western civilization, and why they did so only after the Christian vision of reality took root.[1]

In the ancient world, human value was conditional. Worth was assigned by status, citizenship, gender, productivity, or power. Children could be exposed to die if unwanted. The disabled were considered liabilities. Slaves were property. The poor were burdens. Compassion existed, but it was selective and often transactional. Christianity introduced something quietly radical: the claim that every person bears intrinsic worth because every person is loved by God. Jesus did not teach this as abstract theory. He enacted it. He touched lepers. He welcomed children. He spoke with women publicly. He healed those who could offer nothing in return. He identified Himself with "the least of these," collapsing the distance between divine concern and human vulnerability.[2] Over time, this vision reshaped moral imagination. The idea that dignity is inherent, not earned, became foundational to Western concepts of human rights, even when those rights were later articulated in secular language.

In the Roman world, organized care for the sick and poor was rare. Illness was a private misfortune. Charity was often a means of securing honor or influence. There was no expectation that society as a whole bore responsibility for its most vulnerable members. Early Christians behaved differently. They cared for the sick during plagues, often at great personal risk. They fed the poor not as patrons but as neighbors. They buried the abandoned dead.[3] Over time, this impulse gave rise to institutions that did not previously exist: hospitals, orphanages, hospices—places dedicated to care simply because care was owed. These institutions were not born from efficiency or economic calculation. They emerged from a conviction rooted in the teachings of Christ: that love of neighbor is not optional, and that suffering is never beneath notice.

Ancient societies tended to view power as self-justifying. Kings ruled by strength, divine favor, or inheritance. The idea that rulers were morally accountable to a higher standard was rare and fragile. Christianity altered this calculus by insisting that no human authority is ultimate. Jesus spoke of a kingdom "not of this world" (John 18:36, ESV), yet His teaching did not remove moral pressure from earthly rulers—it intensified it. If all authority is delegated, then all authority can be judged.[4] Over centuries, this conviction contributed to the development of constitutional limits, the rule of law, and the expectation that power must answer to justice rather than merely enforce it. Even secular political systems inherited this assumption: that authority exists to serve, not to dominate.

In many ancient cultures, belief was inseparable from citizenship. To dissent was to betray the state. Christianity introduced a new category into human experience: conscience. Jesus did not coerce belief. He invited it. He allowed rejection. He spoke of truth as something that must be received freely or not at all. This posture eventually gave rise to the idea that belief cannot be compelled without violating the person.[5] Modern concepts of religious liberty and freedom of conscience did not emerge in opposition to Christianity. They grew out of its insistence that faith, to be real, must be freely chosen.

In much of the ancient world, marriage was primarily economic or political. Women were often treated as property, and fidelity was unevenly enforced. Christianity reframed marriage as a mutual moral bond grounded in faithfulness, sacrificial love, and responsibility. Jesus' teaching raised expectations rather than lowered them. Love was no longer merely contractual. It was covenantal.[6] Over time, this vision reshaped norms around consent, fidelity, and the moral equality of men and women within marriage. Changes that later societies would assume without remembering their origin.

Slavery existed across nearly all ancient civilizations and was rarely questioned in moral terms. Christianity did not abolish it overnight. But it planted seeds that would eventually grow into abolition. Paul's letter to Philemon treats a runaway slave as a beloved brother. The early church welcomed slaves and masters as equals before God. Over centuries, Christian theologians developed arguments against slavery grounded in the image of God and natural law.[7] The abolition movement in Britain and America was led overwhelmingly by Christians who appealed explicitly to Scripture and Christian doctrine. William Wilberforce, Harriet Beecher Stowe, Frederick Douglass—these were not secular humanists.

They were believers who saw slavery as a violation of the Christian vision of human dignity.

These transformations did not happen all at once. They were contested. They were incomplete. Christians themselves often failed to see or act on the full implications of their own beliefs. But the direction was clear: societies influenced by Christian teaching moved, however haltingly, toward greater recognition of universal human dignity, moral accountability of power, protection of the vulnerable, and the sanctity of conscience. These were not inevitable developments. They required specific theological convictions about the nature of God, the nature of humanity, and the moral structure of reality.[8]

Consider the development of universities. The first universities in Europe—Bologna, Paris, Oxford—were Christian institutions established to pursue truth within a framework that assumed all truth is God's truth. The conviction that the universe is rationally ordered because it is created by a rational God made systematic inquiry not only possible but morally important.[9] The scientific revolution itself was rooted in this conviction. Figures like Newton, Kepler, Boyle, and Faraday saw their scientific work as uncovering the rationality of the Creator's design. The idea that nature could be understood through mathematical laws presupposed a Lawgiver.

The concept of universal human rights emerged from Christian soil. The medieval theologians of the Salamanca School developed natural law theory, arguing that certain rights belong to all people by virtue of their rational nature, which reflects the image of God.[10] This tradition influenced Locke, whose arguments for natural rights profoundly shaped the American Founding. The Declaration of Independence does not ground rights in government or social contract. It grounds them in the Creator: "We hold these truths to be self-evident, that all men are created equal, that they are endowed by their Creator with certain unalienable Rights."[11] This is not generic theism. It is the Christian understanding that humans, as image-bearers of God, possess inherent dignity and rights that precede and limit government authority.

The separation of church and state, often misunderstood as hostility to religion, also has Christian roots. Jesus' teaching to "render to Caesar the things that are Caesar's, and to God the things that are God's" (Mark 12:17, ESV) introduced a distinction between temporal and spiritual authority that did not exist in the ancient world. Augustine developed this into a theory of two cities: the City of God and the City of Man.[12] This provided the conceptual framework for limiting state power and recognizing spheres of life—family, church, conscience—beyond its reach. The American founders, steeped in this tradition, established a system where government could not establish religion precisely because they took religion seriously. Faith was too important to be coerced or controlled by the state.

What happens when these foundations erode? History provides sobering answers. The totalitarian ideologies of the twentieth century—Nazism, Communism, Fascism—explicitly rejected Christian anthropology. They denied the inherent dignity of persons, the moral accountability of power, and the existence of transcendent truth. The result was catastrophe: gulags, death camps, purges, genocides. These were not aberrations or accidental excesses. They were the logical consequences of worldviews that treated humans as material for ideological projects rather than as image-bearers of God.[13]

Even in less extreme forms, the abandonment of Christian foundations produces moral confusion. If humans are merely evolved animals, why should any individual possess rights? If there is no transcendent standard, why should power be limited? If morality is subjective, why should we condemn injustice? Secular humanism attempts to preserve Christian values without Christian metaphysics, but the experiment has not been successful. Rights become negotiable. Truth becomes relative. Dignity becomes conditional. The moral convictions that once seemed self-evident now require argument, and the arguments often fail to convince because they lack the foundation that originally made them compelling.[14]

This is not a claim that only Christians can be moral or that secular societies are doomed to collapse. Many non-Christians live virtuous lives. Many secular institutions function well. But the question is not whether individuals can be good without God. The question is whether societies can sustain the values of dignity, liberty, and justice without the theological foundations that generated those values. The evidence suggests they cannot. As Western societies drift from Christianity, the concepts it grounded are also drifting. The challenge is not to return to Christendom—that ship has sailed. The challenge is to recognize what we are in danger of losing and to recover the vision of reality that makes those values intelligible and compelling.[15]

Christianity offers what secular alternatives cannot: a coherent account of why every human life matters, why power must be accountable, why truth exists and can be known, why love is more than sentiment, and why justice is more than power. These are not just nice ideas. They are the pillars on which free, flourishing societies rest. And they rest, ultimately, on theological foundations. Remove those foundations, and the pillars will not stand indefinitely on their own.

The heart of Western civilization is not wealth, technology, or military might. It is a set of moral and intellectual commitments rooted in the Christian understanding of God, humanity, and reality. These commitments produced the institutions, values, and practices that made the West what it is. Understanding this is not triumphalism. It is historical accuracy. And recognizing the source of these values is the first step toward preserving them for future generations.

The next section of this book turns inward again, examining what all of this means for you personally. If you live in a universe created by the I AM, if you bear His image, if you are called into relationship with the Triune God, how does this shape your identity, your purpose, and your understanding of what it means to be human?

CHAPTER REVIEW:

KEY CONCEPTS:

- **Christianity's historical contributions** Universities, hospitals, scientific method, human rights, rule of law

- **Gradual moral reforms** Abolition of slavery, protection of vulnerable, women's rights, child labor laws

- **Synthesis of Jerusalem and Athens** Faith and reason, revelation and philosophy working together

- **Secularization without replacement** Modern attempts to keep Christian values while rejecting Christian foundation

- **Current crisis** Erosion of shared moral framework threatens civilizational coherence

WHAT THIS MEANS:

The institutions, values, and practices we take for granted in Western societies did not emerge inevitably but through specific historical development shaped by Christian theology. The secularization project attempts to preserve Christian ethics while discarding Christian metaphysics—a historically unprecedented and likely unsustainable arrangement.

IMPORTANT TERMS & CONCEPTS:

- **Natural Law** (applied to politics) → Chapters 19-20
- **Rule of Law** → Chapter 21 discussion
- **University System** (Christian origins) → Chapter 21 endnotes
- **Secularization** → Chapter 21 analysis
- **Moral Capital** (borrowing without replenishing) → Chapter 21 concept

LOOKING AHEAD:

- Part VII makes this personal living before I AM in contemporary context
  - Q&A Section X addresses living faithfully in secular culture
  - Conclusion (Chapter 23) issues invitation to respond
- This historical analysis supports the book's practical urgency

TECHNICAL REFERENCE:

- For historical documentation: Chapter 21 endnotes (extensive)
- For political philosophy: Chapters 19-21 combined
- For current applications: Q&A Section X (Questions 91-100)
- For civilizational analysis: Introduction; Chapter 21

---

1. For the argument that modern Western values emerged from Christian theology rather than being universal constants discovered by all cultures, see: Holland, T. (2019). *Dominion: How the Christian Revolution Remade the World.* Basic Books. Siedentop, L. (2014). *Inventing the Individual: The Origins of Western Liberalism.* Belknap Press of Harvard University Press. These works trace how concepts of equality, rights, and dignity developed specifically in Christendom.
2. For the Christian transformation of attitudes toward human dignity, especially regarding the vulnerable, see: Stark, R. (1996). *The Rise of Christianity.* Princeton University Press, Chapters 4 and 10. Matthew 25:31-46 (the parable of the sheep and the goats, identifying Christ with the poor and suffering). Early Christian practices of caring for abandoned infants and the sick during plagues contrasted sharply with Greco-Roman norms.
3. For early Christian care during plagues and the development of charitable institutions, see: Stark, R. (1996). *The Rise of Christianity*, Chapter 4 ("Epidemics, Networks, and Conversion"). Harper, K. (2017). *The Fate of Rome: Climate, Disease, and the End of an Empire.* Princeton University Press, pp. 245-248. Christian care during the Antonine and Cyprian plagues (2nd-3rd centuries AD) was distinctive and contributed to Christianity's growth.
4. [4] For the Christian teaching on the accountability of rulers to divine law, see: Romans 13:1-4 (authority from God for justice); Acts 5:29 ("We must obey God rather than men"). Augustine, *City of God*, Book XIX. Aquinas, *Summa Theologica*, I-II, Q. 96, Art. 4 (unjust laws). This tradition provided grounds for resisting tyranny and limiting state power.
5. For the Christian foundations of religious liberty and freedom of conscience, see: Witte, J., Jr. (2000). *Religion and the American Constitutional Experiment*, 2nd edition. Westview Press. Noll, M. A. (2002). *America's God: From Jonathan Edwards to Abraham Lincoln.* Oxford University Press, Chapter 3. The principle that faith cannot be coerced developed in Christian theology long before it was enshrined in law.
6. For the Christian transformation of marriage from contract to covenant, see: Ephesians 5:22-33 (marriage as reflecting Christ and the church). Brundage, J. A. (1987). *Law, Sex, and Christian Society in Medieval Europe.* University of Chicago Press. Christianity introduced consent, mutual fidelity, and permanence as moral requirements for marriage, gradually reshaping legal and cultural norms.
7. [7] For the Christian theological roots of abolitionism, see: Hochschild, A. (2005). *Bury the Chains: Prophets and Rebels in the Fight to Free an Empire's Slaves.* Houghton Mifflin. Wilberforce, W. (1807/1997). *A Letter on the Abolition of the Slave Trade.* Reprinted in various editions. Douglass, F. (1852). "What to the Slave

Is the Fourth of July?" Appeal to Christian conscience. The abolitionist movement was overwhelmingly Christian, grounded in **imago Dei** theology.

8. For the general pattern of Christian influence on Western moral and social development, see: Scrivener, G. (2022). *The Air We Breathe: How We All Came to Believe in Freedom, Kindness, Progress, and Equality.* The Good Book Company. Scrivener argues that modern secular values are borrowed capital from Christianity, often unrecognized by those who hold them.

9. For the Christian foundations of universities and the pursuit of knowledge, see: Pedersen, O. (1997). *The First Universities: Studium Generale and the Origins of University Education in Europe.* Cambridge University Press. Grant, E. (1996). *The Foundations of Modern Science in the Middle Ages.* Cambridge University Press. Medieval universities were ecclesiastical institutions grounded in the conviction that all truth is God's truth.

10. For the Salamanca School's development of natural law and natural rights theory, see: Tierney, B. (1997). *The Idea of Natural Rights: Studies on Natural Rights, Natural Law, and Church Law, 1150–1625.* Eerdmans. Vitoria, F. de (1532/1991). *Political Writings*, edited by A. Pagden and J. Lawrance. Cambridge University Press. The Salamanca theologians developed arguments for universal human rights based on rational nature and the image of God.

11. For the Christian theological foundations of the American Declaration of Independence, see: Dreisbach, D. L., & Hall, M. D., eds. (2014). *The Sacred Rights of Conscience: Selected Readings on Religious Liberty and Church-State Relations in the American Founding.* Liberty Fund. West, T. G. (1997). "The Political Theory of the Declaration of Independence." The Declaration's appeal to "Nature's God" and "Creator" reflects Christian natural theology.

12. For Augustine's two cities doctrine and its influence on church-state relations, see: Augustine. *City of God*, Books XIV–XV, XIX. O'Donovan, O., & O'Donovan, J. L., eds. (1999). *From Irenaeus to Grotius: A Sourcebook in Christian Political Thought.* Eerdmans, pp. 146–207 (Augustine selections). Augustine distinguished temporal and spiritual authority, providing the framework for limiting state power.

13. For the critique of 20th-century totalitarianism as explicitly anti-Christian, see: Solzhenitsyn, A. I. (1973). *The Gulag Archipelago.* Harper & Row. Arendt, H. (1951). *The Origins of Totalitarianism.* Harcourt Brace. Both Nazi and Communist ideologies rejected Christian anthropology, treating humans as material for ideological projects rather than as bearers of divine image.

14. For the argument that secular humanism cannot sustain Christian values without Christian metaphysics, see: MacIntyre, A. (1981). *After Virtue: A Study in Moral Theory.* University of Notre Dame Press, Chapters 1–5. MacIntyre argues that Enlightenment attempts to ground morality in reason alone have failed, leaving modern ethics fragmented and incoherent.

15. For the contemporary challenge of preserving Christian-derived values in post-Christian societies, see: Taylor, C. (2007). *A Secular Age.* Belknap Press of Harvard University Press, especially Part IV. Taylor explores how Western societies have moved from a world where belief in God was the default to one where it is one option among many, and the implications for moral frameworks.

PART SEVEN
THE HUMAN STORY INSIDE GOD'S STORY

CHAPTER TWENTY-TWO
IDENTITY, PURPOSE, AND BEING KNOWN BY I AM

We have traveled from the edges of the cosmos to the foundations of Western civilization, tracing how the God who names Himself "I AM" shapes not only galaxies and laws of nature but also human history, institutions, and moral imagination. These grand themes—the structure of the universe, the nature of being, the moral order embedded in reality—are not abstract or distant. They point toward something profoundly personal. If the eternal I AM created the universe, sustains it moment by moment, and has shaped the course of human civilization, then what does this mean for you? What does it mean to live as someone made in the image of God, called into relationship with the One who spoke all things into being?

Before turning to those deeply personal questions, it is worth pausing to consider one of the most remarkable moments in human history, a moment that bridges the sweep of history with the intimacy of individual life. On a hillside in Galilee, nearly two thousand years ago, Jesus of Nazareth sat down and began to teach. What followed was not a set of legal codes, not a treatise on metaphysics, not a political manifesto. It was something at once simpler and more profound: a description of what human flourishing looks like when lived before the face of God. That teaching has come to be known as the Sermon on the Mount, and it has been recognized, by Christians and non-Christians alike, as one of the most penetrating and beautiful moral visions ever articulated. Leo Tolstoy called it "the basis of all Christian society." Mahatma Gandhi, though not a Christian, said it was the passage in Scripture that most moved him. Even secular ethicists have acknowledged its enduring influence on Western moral thought.[1]

Why does the Sermon on the Mount matter here, at this point in the journey? Because it does what no other teaching does quite so completely: it shows how the grand truths we have been exploring about God, creation, freedom, justice, and love translate into lived reality. The Sermon on the Mount is where cosmic order meets personal identity. It is where the abstract becomes concrete. It is where you discover not only what the universe is, but who you are meant to be within it. And it is the perfect bridge from the historical themes of the previous chapter to the personal questions we now address. The Sermon on the Mount shaped Western civilization precisely because it spoke directly to the human heart, revealing what it means to live as a creature made for relationship with the eternal I AM.[2]

Jesus begins with the Beatitudes, a series of statements that turn conventional wisdom upside down. "Blessed are the poor in spirit, for theirs is the kingdom of heaven. Blessed are those who mourn, for they shall be comforted. Blessed are the meek, for they shall inherit the earth. Blessed are those who hunger and thirst for righteousness, for they shall be satisfied" (Matthew 5:3–6, ESV). These are not platitudes. They are revelations of how God sees human life. To be "poor in spirit" is to recognize that you cannot construct your own meaning, that identity is not self-invented, that you stand before God without pretense or self-sufficiency. It is the acknowledgment that you are a creature, dependent on the One who gives you being. To mourn is not merely to experience sadness but to grieve over brokenness, your own and the world's, and to long for restoration. To be meek is not to be weak but to exercise strength with gentleness, knowing that power is a gift to be stewarded, not a weapon to wield. To hunger and thirst for righteousness is to desire alignment with the moral order God established, to want not just personal success but true goodness, justice, and love.[3]

What is striking about these blessings is how deeply they cut against the human tendency to base identity on self-assertion. The world tells you that significance comes from achievement, status, wealth, or power. The Sermon on the Mount tells you something different: that your deepest identity is not something you construct but something you receive. You are blessed not because of what you accomplish but because of who God is and what He offers. This is the foundation of true identity. You are known by the God who made you. You are loved by the God who sustains you. You are invited into relationship with the God who speaks your name.

The progression of the Beatitudes continues: "Blessed are the merciful, for they shall receive mercy. Blessed are the pure in heart, for they shall see God. Blessed are the peacemakers, for they shall be called sons of God. Blessed are those who are persecuted for righteousness' sake, for theirs is the kingdom of heaven" (Matthew 5:7–10, ESV). These blessings describe what it looks like to live in alignment with God's character. Mercy reflects God's compassion. Purity of heart reflects undivided devotion to truth and goodness. Peacemaking reflects the reconciling work of Christ. Even suffering for righteousness is not meaningless. It reflects participation in the character of God, who does not abandon His people to injustice but stands with them through it.[4]

The Sermon on the Mount also speaks directly to the human experience of being known. Jesus teaches that God knows what you need before you ask (Matthew 6:8, ESV). He knows the desires of your heart, the fears that trouble you at night, the longings you barely dare name. This is not surveillance. It is intimacy. God does not observe you from a distance as though you were an object of curiosity. He knows you as a shepherd knows his sheep, as a father knows his children, as a lover knows his beloved. And His knowledge is rooted in love. He numbers the hairs on your head (Matthew 10:30, ESV). He collects your tears (Psalm 56:8, ESV). Nothing in your life is ignored or forgotten. You matter because the One who sustains existence says you matter.

This truth transforms how you understand yourself. You are not an accident. Your existence is not a cosmic fluke, the random byproduct of impersonal forces grinding through an indifferent universe. You are known. You are named. You are loved. Scripture teaches that God knew you before you were born, that He formed you in the womb, that He has plans for you. Not generic plans for humanity in the abstract, but specific purposes for you (Jeremiah 1:5; Psalm 139:13–16, ESV). The (IAM)illuminates why this makes sense. If the universe is grounded in the personal God who is Being itself, then personhood is not an anomaly. Consciousness is not an illusion. Relationality is not accidental. You were made for relationship with God because the universe itself is structured to support such relationship. The informational architecture of reality, the relational nature of existence, the emergence of minds capable of knowing and loving: all of this points toward the truth that you are meant to know God and be known by Him.[5]

Identity flows from this recognition. You are made in the image of God. This is not a metaphor for your biological form. It is a theological claim about your nature. You are rational, capable of understanding truth. You are moral, capable of discerning right from wrong. You are creative, capable of bringing new things into being. You are relational, capable of love, friendship, and communion. These capacities are not evolutionary accidents. They are reflections of the God in whose image you are made. Your rationality reflects the Logos through whom all things were made. Your moral sense reflects the justice and goodness of God. Your creativity reflects the God who spoke creation into existence. Your capacity for relationship reflects the Triune God, who is eternal communion of Father, Son, and Holy Spirit.

But the Sermon on the Mount does not stop at blessing. It calls for transformation. Jesus teaches that anger is a form of violence, that lust is a form of betrayal, that revenge perpetuates brokenness (Matthew 5:21–48, ESV). He calls His followers to a righteousness that exceeds mere external obedience, a righteousness of the heart. This is not legalism. It is an invitation to become who you were created to be. God does not merely want compliance with rules. He wants the transformation of the person. He wants you to become someone who loves justice not because it is commanded but because it reflects the truth of reality. He wants you to become someone who loves mercy not because it wins approval but because it expresses the character of God. He wants you to become someone who walks humbly not because humility earns rewards but because it acknowledges the reality of creaturehood.[6]

The call to transformation is also a call to purpose. If you are made in the image of God, then your life has direction. You are not adrift in a meaningless universe. You have a role to play in the unfolding of God's purposes. Scripture teaches that you are called to steward creation, to cultivate it, to care for it, to understand it, and to name it. In Genesis, Adam names the animals. An act that acknowledges their natures, their purposes, their places within the order of creation (Genesis 2:19–20, ESV). In the modern world, scientists name stars, classify species, map genomes, and describe laws of nature. Both are acts of understanding. Both reflect the rational capacity that flows from being made in God's image. Your work, your relationships, your creativity, your pursuit of truth—all of this participates in the ongoing actualization of the potentials God placed within creation.

But the deepest purpose is not achievement. It is relationship. You were made to know God and to be known by Him. This is not a distant, impersonal knowledge. It is personal, intimate, and transformative. Jesus calls His followers friends, not servants, because He shares with them the purposes of the Father (John 15:15, ESV). He invites them into communion, into participation in the divine life. The ultimate purpose of your existence is not to accomplish great things, though you may do so. It is not to build a legacy, though you may leave one. It is to love God and be loved by Him, to know Him and be known by Him, to share in the life of the One who is the ground of all being.

The Sermon on the Mount also addresses suffering, and it does so without pretense. Jesus does not promise that those who follow Him will be spared hardship. He promises the opposite: "Blessed are those who are persecuted for righteousness' sake" (Matthew 5:10, ESV). "Blessed are you when others revile you and persecute you and utter all kinds of evil against you falsely on my account" (Matthew 5:11, ESV). This is not masochism. It is realism. The world is broken. Living in alignment with God's character will often place you at odds with systems of power, cultural norms, and human selfishness. But suffering for righteousness is not meaningless. It is participation in the character of God, who does not abandon His people but stands with them through every season of pain, loss, and injustice.

Physics itself points toward this truth in its own language. The universe develops through tension, contrast, and irreversible processes. Complexity arises not from equilibrium but from struggle. Stars form in the violent collapse of gas clouds. Life emerges from chemical systems that consume energy and generate order. Consciousness arises in organisms that navigate danger, solve problems, and adapt to challenges. The arrow of time moves forward not through stasis but through change. In a similar way, the spiritual life deepens through trials. Suffering, when faced with faith, becomes the crucible in which character is formed, wisdom is gained, and compassion is deepened. To be known by God is to be accompanied through every season of life, not exempted from hardship but sustained through it.[7]

The Sermon on the Mount concludes with a parable about foundations. Jesus tells of two builders: one who builds his house on rock, the other on sand. When storms come, the house on rock stands firm, while the house on sand collapses (Matthew 7:24–27, ESV). The point is not merely about wise planning. It is about what your life is built upon. If your identity is built on achievement, status, wealth, or the approval of others, it will not withstand the

Identity, Purpose, and Being Known by I AM

inevitable storms of life. These foundations shift. They fail. They leave you adrift. But if your identity is built on the truth that you are made by God, known by God, and loved by God, then no storm can destroy that foundation. You stand on the bedrock of being itself, on the One who says "I AM."

This brings us back to the questions with which this chapter began. Who are you? You are a creature made in the image of the eternal I AM. You are rational, moral, creative, and relational because God is all of these things. You are known by the One who sustains your existence moment by moment. You are loved by the One who entered history to redeem you. Your identity is not something you construct. It is something you receive.

Why are you here? You are here to know God, to love Him, and to reflect His character in the world. You are here to participate in the ongoing actualization of creation's potentials, to steward the world, to cultivate relationships, to seek truth, to create beauty, to pursue justice, and to show mercy. You are here to become who you were created to be. Someone who lives in alignment with the relational, moral, and rational structure of reality. You are here to bear witness to the One who called you into being.

And what gives your life meaning? Meaning is not self-generated. It is not the product of personal preference or social construction. Meaning flows from being known by the God who is the source of all meaning. Every joy, every sorrow, every triumph, every failure takes on new significance when placed before the One who loves you. Your life is a thread in the vast tapestry of God's unfolding purposes, and every choice you make, every relationship you build, every act of love or justice or mercy contributes to the pattern God is weaving through history.

This is why the Sermon on the Mount is not merely a historical curiosity or a piece of ancient ethical teaching. It is the blueprint for human flourishing. It shows what it means to live as someone made in the image of God, someone who is known and loved by the eternal I AM. It shows that identity is received, not constructed. That purpose flows from relationship with God. That meaning is found not in self-assertion but in self-giving love. And it shows that the moral structure of the universe is not arbitrary or oppressive but the expression of the character of the One who made all things and who invites you into communion with Him.

The next chapter will explore how these truths shape the moral structure of the universe. If God is relational, just, and loving, then the universe must reflect these qualities. If identity and purpose flow from relationship with the I AM, then the way we treat one another, the way we love, the way we pursue justice, the way we navigate freedom must also reflect the character of the One in whose image we are made.[8]

CHAPTER REVIEW:

KEY CONCEPTS:

- **Identity flows from being known by God** Not self-made but received; known before we know ourselves
- **Imago Dei defines true humanity** Rationality, relationality, creativity, moral capacity all reflect divine image
- **Transformation of heart, not just behavior** Sermon on Mount calls for interior righteousness
- **Stewardship vocation** Called to understand, cultivate, and care for creation
- **Relationship as deepest purpose** Made to know and be known by God

WHAT THIS MEANS:

Your identity is not what you achieve or how others see you but who you are before I AM. Being made in God's image gives inherent dignity and defines true purpose: relationship with Creator and faithful presence in creation. The transformation Christianity offers is not religious performance but restored personhood, becoming who you were meant to be.

IMPORTANT TERMS & CONCEPTS:

- **Imago Dei** → Q&A Section III; throughout book
- **Identity (theological)** → Chapter 22 development
- **Vocation** → Chapter 22; theological concept
- **Sanctification** → Part X, Q&A Section VI

Identity, Purpose, and Being Known by I AM

- **Person (as relational)** → Part V; Part X, Section 1.6

LOOKING AHEAD:
- Chapter 23 addresses love and justice in relational universe
- Q&A Section X provides practical guidance for living faith
- Conclusion issues the invitation to respond
- Appendix A provides prayer resources

TECHNICAL REFERENCE:
- For imago Dei comprehensive treatment: Q&A Section III (Questions 21-30)
- For personhood theology: Part V (Chapters 16-18)
- For practical Christian life: Q&A Section VI (Questions 51-60)
- For identity questions: Q&A Section III

---

1. For the historical and ethical significance of the Sermon on the Mount, see: Wright, N. T. (2004). *Matthew for Everyone, Part 1: Chapters 1–15.* Westminster John Knox Press, pp. 31–66. The Sermon on the Mount represents Jesus' most comprehensive ethical teaching and has profoundly shaped Western moral thought. For Gandhi's reflections: Gandhi, M. K. (1940). *An Autobiography: The Story of My Experiments with Truth.* Beacon Press, Part IV, Chapter 19. Gandhi described the Sermon on the Mount as one of the most beautiful things he had ever read.
2. For the influence of the Sermon on the Mount on Western civilization and its embodiment of Christian ethics, see: Stott, J. R. W. (1978). *Christian Counter-Culture: The Message of the Sermon on the Mount.* InterVarsity Press. Bonhoeffer, D. (1959). *The Cost of Discipleship.* Macmillan, Chapters 4–7. These works explore how the Sermon on the Mount presents a radical vision of life before God that has shaped moral imagination across centuries.
3. For theological interpretation of the Beatitudes and their vision of human flourishing, see: Willard, D. (1998). *The Divine Conspiracy: Rediscovering Our Hidden Life in God.* HarperSanFrancisco, Chapters 4–5. Hauerwas, S. (2006). *Matthew.* Brazos Theological Commentary on the Bible. Brazos Press, pp. 63–72. The Beatitudes reveal the upside-down nature of God's kingdom and describe the character formation that accompanies life in communion with God.
4. For the connection between the Beatitudes and participation in God's character, see: Lewis, C. S. (1952). *Mere Christianity.* Macmillan, Book III, Chapters 1–12. Lewis explores how Christian virtues reflect the character of God and how becoming Christlike is the true goal of the moral life. For patristic perspectives: Gregory of Nyssa. *The Beatitudes.* For contemporary treatment: Pinckaers, S. (1995). *The Sources of Christian Ethics.* Catholic University of America Press, Chapter 5.
5. For the theological anthropology underlying human identity as known and loved by God, see: Psalm 139:1–18; Jeremiah 1:5; Romans 8:28–30. Theological

treatment: Barth, K. (1960). *Church Dogmatics* III/2: *The Doctrine of Creation*. T&T Clark, §45 (Man as the Creature of God). Moltmann, J. (1985). *God in Creation: A New Theology of Creation and the Spirit of God*. SCM Press, Chapter 9. Human identity is grounded in being known, loved, and called by God, not in self-construction or social validation.

6. For the Sermon on the Mount's call to transformation rather than mere external obedience, see: Matthew 5:21–48. Theological exposition: Calvin, J. *Commentary on a Harmony of the Evangelists, Matthew, Mark, and Luke,* Volume 1 (on Matthew 5–7). For contemporary treatment: Pennington, J. T. (2007). *The Sermon on the Mount and Human Flourishing: A Theological Commentary*. Baker Academic. Jesus calls for righteousness that exceeds scribal obedience by addressing the heart, not just behavior.

7. For suffering as participation in God's purposes and the formation of character, see: Romans 5:3–5; James 1:2–4; 1 Peter 1:6–7. Theological reflection: Lewis, C. S. (1940). *The Problem of Pain*. Macmillan, Chapters 6–7. Plantinga, C. (1995). *Not the Way It's Supposed to Be: A Breviary of Sin*. Eerdmans, Chapter 3. Suffering, though not good in itself, can be redeemed and become the means of spiritual growth when faced with faith.

8. For the parable of the two builders and its implications for grounding life in God, see: Matthew 7:24–27. Theological commentary: Augustine. *Sermon on the Mount* (Books I and II). For contemporary application: Peterson, E. H. (2005). *Christ Plays in Ten Thousand Places: A Conversation in Spiritual Theology*. Eerdmans, Part I. Building one's life on the words of Christ means grounding identity, purpose, and meaning in the unchanging character of God rather than shifting circumstances.

CHAPTER TWENTY-THREE
LOVE, JUSTICE, AND THE SHAPE OF A RELATIONAL UNIVERSE

*I*f you are made in the image of God, known by Him, and called into relationship with Him, then what does this mean for how you live? What does it mean for how you treat others? What does it mean for the choices you make, the relationships you build, and the world you help shape? The previous chapter explored identity and purpose as gifts received from the eternal I AM. This chapter turns to the outworking of those truths in daily life. If the universe is grounded in the God who is love, justice, and faithfulness, then the moral structure of reality must reflect these qualities. And if you are made in God's image, then your life must reflect them as well.

The Sermon on the Mount does not remain abstract. After describing the blessings of those who live before God, Jesus moves to practical application. He calls His followers to be salt and light, to let their good works shine before others so that people may see and give glory to God (Matthew 5:13–16, ESV). This is not showmanship. It is witness. Your life is meant to reflect the character of the One who made you. When you act with love, you reveal something about the nature of God. When you pursue justice, you participate in the moral order He established. When you show mercy, you reflect His compassion. The moral life is not external obedience to arbitrary rules. It is participation in the very structure of reality, alignment with the character of God, and cooperation with the purposes for which you were made.[1]

Love is the highest expression of God's nature. Scripture declares that "God is love" (1 John 4:8, ESV). This does not mean that love is one attribute God possesses among many, like kindness or power. It means that love is essential to who God is. The Trinity is eternal love—Father, Son, and Holy Spirit—giving and receiving in perfect communion. Love, in its essence, is the willing of the good for another. It is relational, self-giving, and creative. It does not seek its own but delights in the flourishing of the beloved. And because God is love, the universe He creates is structured for relationship. You are capable of love because you are made in the image of the God who is love. Your capacity to care for another person, to sacrifice your own comfort for their well-being, to delight in their joy and grieve at their suffering. All of this reflects the relational nature of the One who made you.[2]

But love is not sentimentality. It is not mere affection, though affection may accompany it. Love is the commitment to the good of the other, even when that commitment is costly. Jesus teaches this in the Sermon on the Mount when He calls His followers to love not only their friends but also their enemies. "You have heard that it was said, 'You shall love your neighbor and hate your enemy.' But I say to you, Love your enemies and pray for those who persecute you, so that you may be sons of your Father who is in heaven" (Matthew 5:43–45, ESV). This is not natural. It is supernatural. It is the extension of divine love into human relationships. To love your enemy is not to ignore the harm they have caused or to pretend that evil is good. It is to desire their redemption, to refuse to be consumed by hatred, and to trust that God's justice will address what you cannot.[3]

It is often argued that love requires freedom. A relationship without the possibility of refusal is not a relationship. It is compulsion, control, or programming. If God desires communion with His creatures. If He calls them not to be servants but friends, not puppets but persons. Then He must endow them with free will. Freedom is not a flaw in creation or an unfortunate necessity. It is a gift necessary for the fulfillment of love. Without freedom, there can be no genuine relationship. Without genuine relationship, there can be no love. This is why the universe exhibits openness, indeterminacy, and genuine choice. Quantum events are not predetermined. Biological systems adapt and vary. Human decisions involve real deliberation. This openness is not chaos. It is the space in which love can flourish, in which persons can grow, and in which freedom can be exercised toward the good.[4]

The relational nature of creation reflects this truth. You are not an isolated atom operating in mechanical independence. You are a person embedded in relationships, capable of genuine agency within a world that responds to your choices. Every decision you make ripples through the web of relationships that constitutes human life. You affect others. You shape communities. You leave marks on history. This is what it means to be free in a relational universe. And this is why your choices matter. The universe is structured in such a way that love, justice, and mercy have real consequences. Acts of kindness strengthen the fabric of reality. Acts of cruelty tear it. Acts of truth increase clarity. Acts of deception generate confusion. The moral structure of the universe is not imposed arbitrarily from outside. It is woven into the nature of reality itself.

Justice is the second pillar of the moral universe. Justice is the right ordering of relationships. It ensures that the dignity of each person is honored, that wrongs are addressed, and that the good is protected. In Scripture, justice is not merely legal or procedural. It is moral, relational, and rooted in the character of God. God is just because God is faithful to Himself, to His creation, and to the moral order He established. Injustice is a violation of reality because it contradicts the nature of the God who is truth. The Sermon on the Mount addresses justice repeatedly. Jesus calls His followers to reconcile with those they have wronged before offering worship

(Matthew 5:23–24, ESV). He teaches that harboring anger or contempt is itself a form of violence that damages relationships and disrupts the moral order (Matthew 5:21–22, ESV). He warns against using religious language to mask injustice or to justify self-righteousness (Matthew 6:1–18, ESV). Justice, in the vision of the Sermon on the Mount, is not about winning arguments or asserting rights. It is about restoring what has been broken, reconciling what has been estranged, and ensuring that relationships reflect the truth of who we are as image-bearers of God.[5]

The universe itself exhibits this relational justice. Actions reverberate through the fabric of reality. Harm to others disrupts the relational order. Lies create distrust. Violence breeds fear. Exploitation generates oppression. Conversely, fidelity strengthens bonds. Compassion heals wounds. Honesty builds trust. Truth contributes to flourishing. Justice is not simply imposed from above by divine decree. It is woven into the structure of creation. Scripture calls this the law written on the heart (Romans 2:14–15, ESV), and the classical tradition describes it as natural law: the participation of rational creatures in the eternal law of God. When you act justly, you align yourself with the way things truly are. When you act unjustly, you act against the grain of reality, and the consequences follow not as arbitrary punishment but as the natural result of violating the relational structure of existence.[6]

Love and justice are not opposites. They are complementary aspects of God's character and must be held together in any rightly ordered life or society. Love without justice is sentimentality that tolerates evil and fails to protect the vulnerable. Justice without love is cruelty that insists on punishment without mercy or restoration. God holds them together perfectly. His commands are expressions of His love, protecting what is good and guiding us toward flourishing. His judgments are expressions of His justice, addressing wrongs and vindicating the oppressed. You are called to imitate this unity. The Sermon on the Mount commands both: "Blessed are the merciful, for they shall receive mercy" (Matthew 5:7, ESV), and "Blessed are those who hunger and thirst for righteousness, for they shall be satisfied" (Matthew 5:6, ESV). Mercy and righteousness, compassion and justice—these belong together. To love mercy is to extend grace, forgiveness, and kindness even when they are not deserved. To do justice is to uphold what is right, to protect the weak, and to hold the powerful accountable. Both are required. Both reflect the character of God.[7]

The universe supports this calling because it is relational from the ground up. Every act of love strengthens the fabric of reality. Every act of justice restores what has been damaged. Every choice to honor truth, show compassion, or defend the vulnerable participates in the relational logic that sustains creation. The (IAM) clarifies this vision. The universe grows through interactions, distinctions, and relationships. Human life mirrors this process. Every act of truth increases clarity. Every act of compassion strengthens connection. Every act of justice restores balance. The moral life is not merely obedience to external rules. It is participation in the relational logic of creation. It is alignment with the character of God. It is cooperation with the purposes woven into the fabric of reality.

This brings us to the problem of evil. If the universe is grounded in the I AM who is good, why does evil exist? This is one of the oldest and most pressing questions in theology and philosophy. The Sermon on the Mount does not shy away from it. Jesus acknowledges that His followers will face persecution, suffering, and injustice (Matthew 5:10–12, ESV). He does not promise exemption from hardship. He promises presence through it. The answer to the problem of evil lies in the nature of freedom. God creates beings capable of genuine relationship, which requires genuine freedom. Evil is the misuse of this freedom. It is the rejection of the relational order God established. It is the choice to turn away from the good, to act against truth, to harm rather than heal.[8]

Evil is real. It causes suffering. It disrupts relationships. It leaves scars on individuals and societies. But evil does not negate the goodness of God. Instead, it reveals the seriousness of freedom and the weight of human choices. God permits evil not because He is indifferent or powerless, but because He desires love, and love cannot exist without the possibility of refusal. To eliminate the possibility of evil would require eliminating freedom. To eliminate freedom would require eliminating persons. And a universe of mechanisms, however orderly, could never fulfill the purpose for which God creates: communion with beings capable of knowing and loving Him freely.

This does not make evil good or excuse its presence. It explains why a good God would create a world in which evil is possible. And it places the responsibility for evil where it belongs: not with the Creator, but with the creatures who misuse their freedom. God provides the ground of being. He establishes the moral order. He invites us into relationship. But He does not force compliance. We choose. And when we choose wrongly, the consequences follow not as divine punishment arbitrarily imposed but as the natural result of acting against the structure of reality.

Yet Scripture teaches that God does not abandon creation to the consequences of evil. He works within history to restore, redeem, and heal. The incarnation is the ultimate expression of this restoration. The Logos, through whom all things were made, enters the world He created. He takes on human flesh, experiences human suffering, confronts evil directly, and offers the path of redemption. The cross is where love and justice meet. Justice is satisfied because sin is addressed, its consequences borne by the sinless one. Love is fulfilled because God gives Himself for the sake of His creatures, absorbing the cost of rebellion in order to offer forgiveness and new life. This is not a legal transaction detached from relationship. It is the climax of God's relational commitment to His creation.[9]

The resurrection is the promise that the relational order will ultimately be restored. Death, the final consequence of sin and the ultimate disruption of relationship, is defeated. The body is not discarded but transformed. Creation is not abandoned but renewed. The trajectory of history is not circular or meaningless but directed toward the restoration of all things, the reconciliation of heaven and earth, and the full realization of God's purposes for

His creation. Evil is real, but it is not ultimate. Suffering is present, but it is not permanent. The brokenness we experience now is not the final word. God's love is stronger than death, and His justice will prevail.

The moral structure of the universe also reveals human responsibility. Every person is called to participate in the good, to seek justice, to love mercy, and to walk humbly with God (Micah 6:8, ESV). The relational nature of reality means that every choice affects others. No one lives in isolation. Human life is a tapestry of relationships, and each decision strengthens or weakens the fabric of that tapestry. The call to love and justice is not a burden imposed from outside. It is the path to human flourishing and the fulfillment of your nature. You were made for relationship. You were made for love. You were made to reflect the character of the God in whose image you are created. And when you live in accordance with that purpose, you experience the fullness of what it means to be human.

The Sermon on the Mount concludes with a call to action: "Everyone then who hears these words of mine and does them will be like a wise man who built his house on the rock" (Matthew 7:24, ESV). Hearing is not enough. Understanding is not enough. The moral vision Jesus presents must be lived. It must shape your relationships, your choices, your commitments, and your character. And when it does, you discover that the moral order is not a set of arbitrary constraints but the pathway to life. Justice protects dignity. Love creates flourishing. Mercy heals wounds. Truth sets free. These are not abstract ideals. They are the practical outworking of living in alignment with the God who is the ground of all being.

This vision integrates what we have learned from physics, philosophy, and theology. Physics reveals a universe that is ordered, relational, and open to novelty. Philosophy shows that this order requires a ground, that relationality points toward the Trinity, and that freedom is the condition for personhood. Theology identifies the personal God who is the source of all being, who creates out of love, who sustains through His Word, and who redeems through the incarnation. Together, these perspectives show that the universe is not a brute fact, not an accident, not a meaningless collection of particles. It is the expression of divine love, the arena of genuine freedom, and the stage on which the drama of redemption unfolds.

Love requires freedom. Justice preserves dignity. Evil arises from the misuse of freedom but does not negate God's goodness. Redemption is offered through Christ. And the moral life is participation in the relational structure of reality itself. These truths are not abstractions. They are the foundations of meaningful existence. They explain why your choices matter, why relationships are precious, why justice is worth pursuing, and why hope is justified even in the face of suffering.

The next section of this book turns to the Master Q&A, where the deepest questions raised throughout these chapters will be addressed directly and clearly. Each answer draws on theological, philosophical, and scientific insight, providing Scripture references for further understanding and offering practical guidance for living in light of these truths. The questions cover objections, clarifications, and applications, ensuring that the vision presented here is not merely theoretical but lived, not merely intellectual but transformative.[10]

CHAPTER REVIEW:

KEY CONCEPTS:

- **Love as fundamental to reality** Because God IS love (1 John 4:8), not just has love
- **Justice as right ordering of relationships** Giving each person their due; reflecting divine character
- **Moral law reflects relational structure** Universe built for communion, not conflict
- **Integration of love and justice** Not competing values but complementary expressions of God's nature
- **Invitation to participation** Living in alignment with reality's relational grain

WHAT THIS MEANS:

Love is not sentiment but the fundamental structure of reality reflecting the Triune God. Justice is not arbitrary rules but right relationship. Living according to love and justice is not restriction but alignment with how things actually are. The universe is built for this; we're called to live accordingly.

IMPORTANT TERMS & CONCEPTS:
- **Love (theological definition)** → Part X, Section 1.1 (Theology)
- **Justice (philosophical/theological)** → Chapter 20; Part X concepts
- **Relationality** → Throughout book; Part X, Section 7.7
- **Moral Law** → Chapter 20; Q&A sections
- **Participation** → Part X, Section 1.6 (Participation)

LOOKING AHEAD:
- Conclusion (next) issues personal invitation to respond
- Q&A Section X addresses practical living questions
- Appendix A provides resources for response
- This completes the book's argument before the invitation

TECHNICAL REFERENCE:
- For love as divine essence: Part X, Q&A Section I; Part V
- For justice in three domains: Chapter 20; Part X, Section 7
- For practical application: Q&A Section X (Questions 91-100)
- For relational ontology: Part V; Part X, Section 7.7

---

1. For the call to be salt and light as participation in God's character, see: Matthew 5:13–16. Wright, N. T. (2004). *Matthew for Everyone, Part 1: Chapters 1–15*. Westminster John Knox Press, pp. 36–40. The metaphors of salt and light emphasize both preservation and illumination—Christians are called to preserve what is good and to illuminate truth through their actions.
2. For the theology of God as love and the Trinitarian nature of divine love, see: 1 John 4:8, 16; John 17:20–26. Moltmann, J. (1981). *The Trinity and the Kingdom*. SCM Press, Chapters 1–2. Zizioulas, J. D. (1985). *Being as Communion: Studies in Personhood and the Church*. St Vladimir's Seminary Press. Love is not an attribute God possesses but the essence of who God is as eternal communion.
3. For Jesus' teaching on love of enemies and its radical nature, see: Matthew 5:43–48; Luke 6:27–36. Theological treatment: Yoder, J. H. (1972). *The Politics of Jesus*. Eerdmans, Chapter 6. Bonhoeffer, D. (1959). *The Cost of Discipleship*. Macmillan, Chapter 8. Love of enemies is the most distinctive element of Christian ethics,

reflecting the character of God who "makes his sun rise on the evil and on the good" (Matthew 5:45).

4. For the argument that love requires freedom, see: Swinburne, R. (1998). *Providence and the Problem of Evil*. Oxford University Press, Chapters 5–6. Plantinga, A. (1974). *The Nature of Necessity*. Oxford University Press, Chapter 9 (the free will defense). A relationship without genuine freedom is not a relationship but compulsion; God creates free beings capable of love because love cannot be coerced.

5. For biblical teaching on justice and reconciliation, see: Matthew 5:21–26; Micah 6:8; Amos 5:21–24. Wolterstorff, N. (2008). *Justice: Rights and Wrongs*. Princeton University Press. Cone, J. H. (1975). *God of the Oppressed*. Seabury Press. Justice in Scripture is restorative and relational, concerned with right relationships and the protection of the vulnerable.

6. For the concept of natural law as God's moral order written into creation and conscience, see: Romans 2:14–15; Aquinas, T. *Summa Theologica*, I-II, Questions 90–94. Budziszewski, J. (2003). *What We Can't Not Know: A Guide*. Spence Publishing. Natural law affirms that certain moral truths are accessible to reason and written on the heart, reflecting God's eternal law.

7. For the integration of love and justice in God's character and Christian ethics, see: Psalm 89:14 ("Righteousness and justice are the foundation of your throne; steadfast love and faithfulness go before you"). Volf, M. (1996). *Exclusion and Embrace: A Theological Exploration of Identity, Otherness, and Reconciliation*. Abingdon Press. Both mercy and justice flow from God's nature and must be held together.

8. For the free will defense and the problem of evil, see: Plantinga, A. (1974). *God, Freedom, and Evil*. Eerdmans. Hick, J. (1966). *Evil and the God of Love*. Macmillan (Irenaean theodicy). These works argue that the possibility of moral evil is a necessary consequence of creating beings with genuine freedom capable of love and moral responsibility.

9. [9] For the cross as the meeting of love and justice, see: Romans 3:21–26; 2 Corinthians 5:21. Stott, J. R. W. (1986). *The Cross of Christ*. InterVarsity Press, Chapter 5. Torrance, T. F. (1992). *The Mediation of Christ*, revised edition. Helmers & Howard, Chapters 2–3. At the cross, God's justice is satisfied and His love is displayed, providing both atonement for sin and reconciliation with God.

10. For the practical application of the Sermon on the Mount to daily life, see: Bonhoeffer, D. (1959). *The Cost of Discipleship*. Macmillan, Part I. Willard, D. (1998). *The Divine Conspiracy: Rediscovering Our Hidden Life in God*. HarperSanFrancisco, Chapters 4–9. The Sermon on the Mount is not merely ethical teaching but a description of life in God's kingdom, meant to be lived rather than merely admired.

PART EIGHT
THE MASTER Q&A SECTION

COMPREHENSIVE QUESTIONS ON PHYSICS, PHILOSOPHY, THEOLOGY, AND THE NATURE OF I AM

This section contains 100 questions organized into 10 subsections. Each section provides clear, direct answers to the major questions raised throughout the book. The answers integrate modern physics and metaphysics with biblical theology. Scripture references appear naturally within the text to guide further study.

SECTION 1

GOD AND THE NATURE OF I AM

(Questions 1-10)

1. Does God know our thoughts?

Yes. Scripture teaches that God knows the inner life of every person. *Psalm 139* says that God searches us and knows us, that He discerns our thoughts from afar, and that every word is known before it is spoken. *Hebrews 4* explains that all things are laid bare before the eyes of Him to whom we must give account. God knows our thoughts not through observation but because our very being depends on Him. His knowledge is relational, personal, and grounded in His intimate presence. *References: Psalm 139:1–4; Hebrews 4:13; Jeremiah 17:10.*

2. How can God be both transcendent and immanent?

God is transcendent because He exists beyond creation. He is immanent because He is present within it. Scripture teaches both truths. Isaiah declares God as the One who inhabits eternity, while Paul teaches that in Him we live and move and have our being. The name "I AM" captures this dual reality. God is not a distant deity nor a force trapped inside the universe. He is the ground of being who gives existence to all things while remaining distinct from them. *References: Isaiah 57:15; Psalm 139:7–10; Acts 17:28.*

3. Why did God reveal Himself as "I AM That I AM"?

The divine name reveals that God is the source of all being. When God says, "I AM," He is not describing a characteristic but His essence. God is existence itself. This stands in stark contrast to every created thing, whose essence does not guarantee its existence. A tree might exist or might not exist; its nature as "tree" doesn't require its existence. But God's very nature *is* to exist. He cannot not-be.

The name also emphasizes God's faithfulness and unchanging presence. The Hebrew **Ehyeh Asher Ehyeh** can be translated "I AM WHO I AM" or "I WILL BE WHO I WILL BE." Both convey the same truth: God is eternally present, absolutely reliable, and completely self-sufficient. He depends on nothing outside Himself. All creation depends on Him.

This name was given to Moses at a critical moment—when Israel needed deliverance from Egypt. God revealed Himself not as a distant philosophical principle but as the living God who acts in history, hears the cries of His people, and intervenes on their behalf. "I AM" is the God who was with the patriarchs, who is with Moses now, and who will be with Israel in the future.

The significance of this name echoes throughout Scripture. Jesus applies it to Himself in John's Gospel: "Before Abraham was, I AM " *(John 8:58, ESV)*. In Revelation, God declares, "I AM the Alpha and the Omega...who is and who was and who is to come" *(Revelation 1:8, ESV)*. The name "I AM " reveals the eternal, self-existent, unchanging God who is the ground of all reality. *References: Exodus 3:14; Revelation 1:8; John 8:58; Isaiah 41:4.*

4. How does the Trinity relate to the structure of the universe?

The Trinity reveals that God is relational in His very essence. Father, Son, and Spirit are distinct persons but one God. This means that relationality is not an afterthought or a human projection. It is woven into ultimate reality. Physics shows that the universe is relational at its foundation: particles are entangled, information is shared, and nothing exists in isolation. The Trinity provides the metaphysical explanation for this relational universe. Creation reflects the Creator. *References: John 1:1–3; John 14:16–20; Matthew 28:19.*

5. If God is all-powerful, why does He work through processes?

God often works through processes because He created a world in which development reveals His wisdom. Scripture shows God creating, forming, guiding, calling, and fulfilling. He works through seasons, history, and growth. Physics also reveals a universe that unfolds gradually: stars form, galaxies assemble, life emerges, and consciousness grows. God's power is not diminished by this. Processes display His wisdom, patience, and relational character. *References: Genesis 1; Ecclesiastes 3:1–11; Galatians 4:4.*

6. Is God affected by time?

No. God is eternal and does not experience time as creatures do. Time is a property of creation that emerges through change and entropy. Scripture teaches that God is the One who was, who is, and who is to come, not because He is locked in sequence but because He transcends it. All moments are present to Him. God interacts with creation in time, but His nature is timeless. *References: Psalm 90:4; 2 Peter 3:8; Isaiah 57:15.*

7. If God is unchanging, how can He respond to prayer?

God responds to prayer from His eternal vantage. His unchanging nature does not prevent relational interaction. Instead, His unchanging love ensures He always responds perfectly. Scripture portrays God as hearing, answering, and engaging with His people. From the standpoint of eternity, God encompasses every moment of prayer and every moment of His response simultaneously. Prayer does not change God's nature; it draws us into participation with His will. *References: Psalm 34:17; Matthew 7:7–11; James 5:16.*

8. How can a personal God create an impersonal universe of laws?

The universe is not impersonal. It is ordered, intelligible, and relational. Laws do not remove God's presence; they express His rationality. Scripture teaches that God upholds all things by His word. The laws of nature are stable because God is faithful. They are intelligible because the Logos orders all things. Physical law manifests divine rationality, not its negation. *References: Hebrews 1:3; Psalm 119:89–91; John 1:3.*

9. Does God limit Himself?

God does not limit His nature, but He freely chooses to act in ways consistent with His character. Scripture teaches that God cannot lie, deny Himself, or act unjustly. These are not limitations but expressions of His perfection. God also allows human freedom and the integrity of creation. These too are voluntary acts flowing from His relational and loving nature. *References: 2 Timothy 2:13; Titus 1:2; James 1:13.*

10. Why did God create anything at all?

Scripture teaches that God created out of love, not out of need. God is complete within Himself. Creation is the overflow of His goodness. The universe reflects His glory, His wisdom, and His relational nature. A relational God naturally creates a relational universe, and a rational God creates a rational universe. Existence itself is an expression of divine generosity. *References: Psalm 19:1; Revelation 4:11; Isaiah 43:7.*

SECTION II

CREATION, PHYSICS, AND THE UNIVERSE

(Questions 11–20)

11. How did God create the universe out of nothing?

"Scripture teaches that God created the universe by His Word, not from preexisting material. Creation out of nothing means that all matter, energy, space, time, and physical law came into being by divine command. Physics supports this remarkably well: the universe has a finite beginning, and time itself began at the Big Bang. Classical theology describes creation as the transition from pure potentiality in the divine mind into actualized existence within spacetime. The universe exists because God wills it." *References: Genesis 1:1–3; Hebrews 11:3; Psalm 33:6.*

12. Does the Big Bang contradict creation?

No. The Big Bang affirms that the universe had a beginning, which is precisely what Scripture teaches. The idea that time, space, and matter suddenly came into existence aligns with the biblical declaration that God spoke creation into being. The Big Bang describes how the universe began to unfold, but not why it exists. The cause of the beginning must lie outside time and matter, which is consistent with the God who says "I AM." *References: Genesis 1:1; John 1:1–3.*

13. What existed before the universe began?

Before the universe began, God alone existed. Time, space, and physical law had no reality apart from God's creative will. Scripture teaches that God inhabits eternity and that all things were created through Him. Physics cannot describe a "before" the Big Bang because time itself did not exist. The eternal God is not bound by temporal sequence. He exists beyond time and brings time into being. *References: Psalm 90:2; Isaiah 57:15; Colossians 1:16–17.*

14. Is the universe finely tuned for life?

Yes. The physical constants of the universe are set with extraordinary precision. The strength of gravity, the charge of the electron, the cosmological constant, and dozens of other parameters fall within narrow ranges that allow life to exist. Scripture teaches that God created the earth to be inhabited and that the heavens declare His glory. Fine-tuning reflects intentional design. The relational architecture of the universe is consistent with a Creator who orders reality through wisdom. *References: Isaiah 45:18; Psalm 19:1.*

15. What does modern physics reveal about how God creates?

Modern physics shows that the universe is fundamentally informational, relational, and mathematical. At the quantum level, particles are not solid objects but patterns of information and probability. The structure of reality emerges from relationships and interactions rather than from isolated, independent matter.

This aligns remarkably with Scripture's account of creation. The Bible teaches that God creates through His Word: "By the word of the LORD the heavens were made" *(Psalm 33:6, ESV).* When God speaks, information becomes reality. His command brings order out of chaos, structure out of formlessness. The physical world is not generated from pre-existing matter but from the immaterial—the divine intention expressed as Word.

Hebrews affirms this: "By faith we understand that the universe was ordered by the word of God, so that what is visible came into being through the invisible" (Hebrews 11:3, NABRE). The visible material world emerges from the invisible—from information, pattern, and divine intention. Physics now describes this same reality from its own perspective: matter is not ultimately "stuff" but structured information.

The holographic principle in physics demonstrates that information about a region can be encoded on its boundary. Black-hole thermodynamics shows that entropy (information content) scales with surface area rather than volume. These discoveries suggest that reality itself is information-theoretic at its foundation—precisely what we would expect if the universe were spoken into existence by a rational God.

This does not prove God exists. Physics describes **how** the universe operates; theology explains **why** it exists and **who** sustains it. But the informational nature of reality is deeply consonant with the biblical teaching that creation flows from the divine Logos—the Word who is both the rationality of God and the agent of creation (John 1:1-3). *References: Psalm 33:6, 9; Hebrews 11:3; John 1:1-3; Colossians 1:16-17; Genesis 1:3.*

16. If physics says reality is informational, does that mean matter isn't real?

No. Matter is genuinely real, but its deepest nature is not what common sense suggests. Modern physics reveals that what we experience as solid matter is actually structured energy and information. Particles are excitations in quantum fields. Their properties—mass, charge, spin—emerge from mathematical relationships and symmetries. Reality is relational and informational at its foundation, but this does not make it less real.

Scripture anticipated this insight. Paul writes that "what is seen was not made out of things that are visible" (*Hebrews 11:3, ESV*). The physical world we observe emerges from something immaterial—from God's creative Word, which is itself informational. The visible depends on the invisible. This is not illusion but dependence. The material world is genuine because it is grounded in God's creative intention.

Colossians teaches that "in him all things were created, in heaven and on earth, visible and invisible" and that "in him all things hold together" (*Colossians 1:16-17, ESV*). Christ is the sustaining principle of the universe. Physical laws are stable because God is faithful. Matter coheres because the Word upholds it. The informational structure of reality points to the Logos through whom all things were made.

The fact that matter is informational does not reduce it to illusion. Information is real. Relationships are real. Patterns are real. What physics has discovered is that the **nature** of material reality is more subtle, more relational, and more dependent on mind and meaning than materialism assumed. This makes the universe more, not less, compatible with the Christian claim that it is the product of divine reason. *References: Hebrews 11:3; Colossians 1:16-17; John 1:3; Psalm 104:30; Acts 17:28.*

17. Does quantum mechanics point toward God?

Quantum mechanics reveals that reality is irreducibly relational. Particles do not have definite properties until they interact. Measurement does not simply reveal pre-existing facts; it participates in actualizing specific outcomes from potentiality. The universe at its foundation is not a collection of isolated, independent objects but a web of relationships and interactions.

This does not prove God exists, but it decisively undermines the materialist worldview that dominated science for centuries. Materialism assumed the universe was made of solid, independent particles governed by deterministic laws, with consciousness being an accidental byproduct. Quantum mechanics shows that reality cannot be described that way. The observer and the observed are intertwined. Reality is participatory, not mechanistic.

Scripture has always taught that creation is relational and sustained. Paul declares that "in him we live and move and have our being" (*Acts 17:28, ESV*). The author of Hebrews writes that Christ is "upholding the universe by the word of his power" (*Hebrews 1:3, ESV*). These are not ancient myths but statements about the nature of reality: existence is not self-sustaining but continuously dependent on God.

Quantum mechanics also reveals that the universe is rational and mathematical, governed by elegant equations and deep symmetries. This aligns with the biblical teaching that creation flows from divine Wisdom and Logos. Why should the universe be comprehensible to the human mind unless both mind and universe share a common rational source?

Physics does not replace theology. It operates within its proper domain, describing regularities and mechanisms. But the kind of universe quantum mechanics reveals—relational, rational, participatory, and intelligible—is precisely the kind of universe we would expect if it were grounded in a personal, rational Creator. *References: Acts 17:28; Hebrews 1:3; Colossians 1:17; Psalm 104:24; Proverbs 8:22-31; John 1:3.*

18. Why does time move in one direction?

Scripture presents time as purposeful and directional. History is not cyclical but linear—it moves from creation toward consummation, from Genesis to Revelation, from promise to fulfillment. God is the Alpha and Omega, the beginning and the end, yet He acts within time, entering history at specific moments: calling Abraham, delivering Israel, becoming incarnate in Christ, and promising to return.

Physics describes time's direction through the increase of entropy—the tendency of systems to move from order toward disorder, from low entropy toward higher entropy. This thermodynamic arrow of time gives us past and future, cause and effect, memory and anticipation. The universe began in a highly ordered state at the Big Bang and has been evolving toward greater complexity and entropy ever since.

Ecclesiastes teaches that there is "a time for every matter under heaven" (*Ecclesiastes 3:1, ESV*). Time is the medium in which God's purposes unfold. Creation itself is temporal; it has a beginning and will have an end. The flow of time is not arbitrary but purposeful. It allows for growth, development, choice, relationship, and narrative.

A timeless universe would be static. Time is the gift that makes history, and therefore redemption, possible.

The Bible also teaches that God transcends time. For God, "a thousand years are as a day" (*2 Peter 3:8*). He is not bound by temporal succession but sees all moments at once from the perspective of eternity. Yet He chooses to work within time for our sake. The incarnation is the supreme example: "When the fullness of time had come, God sent forth his Son" (*Galatians 4:4, ESV*). Time is real, meaningful, and purposed.

Physics and Scripture agree: time is not an illusion. It has direction. It has meaning. The arrow of time points toward the consummation of all things when Christ returns and God makes all things new References: *Genesis 1:1; (Revelation 21:5).Revelation 22:13; Ecclesiastes 3:1-11; 2 Peter 3:8; Galatians 4:4; Hebrews 1:2.*

19. What does the Bible teach about the structure of the universe?

Scripture does not provide a scientific cosmology, nor should we expect it to. The Bible's purpose is to reveal God and His relationship with humanity, not to offer technical descriptions of astrophysics. Yet the Bible does make theological claims about the universe that have profound implications.

First, Scripture teaches that the universe is created, not eternal. "In the beginning, God created the heavens and the earth" (*Genesis 1:1, ESV*). This stands against ancient pagan cosmologies that assumed eternal matter or cyclical time. Modern cosmology confirms that the universe had a beginning—the Big Bang marks the origin of space, time, and matter.

Second, Scripture teaches that the universe is ordered and intelligible. God establishes "ordinances of the heavens" (*Job 38:33, ESV*). The regularities of nature are not accidents but reflections of divine wisdom. *Psalm 19* declares, "The heavens declare the glory of God, and the sky above proclaims his handiwork" (*Psalm 19:1, ESV*). The rational order of the cosmos points beyond itself to the mind of the Creator.

Third, Scripture teaches that the universe is finite and bounded.

Isaiah writes that God "stretches out the heavens like a curtain, and spreads them like a tent to dwell in" (*Isaiah 40:22, ESV*). This imagery suggests expansion and boundary, which aligns with modern cosmology's description of an expanding universe with a finite observable region.

Fourth, Scripture teaches that the universe is sustained by God's word. It is not self-sufficient. Christ is "upholding the universe by the word of his power" (*Hebrews 1:3, ESV*). Without God's continuous sustaining action, creation would cease to exist. The physical laws are stable because God is faithful.

Modern physics has proposed many models—some universe as a holographic projection, the multiverse hypothesis, or the mathematical structure of spacetime. These remain speculative in many respects. What matters biblically is not the precise technical details but the theological truths: the universe is created, ordered, finite, contingent, and dependent on God. Whatever the scientific model, these truths remain.

The cosmos is not divine. It is not eternal. It is not self-explanatory. It is creation—finite, beautiful, rational, and good—pointing always to the infinite God who made it and holds it in being. *References: Genesis 1:1; Psalm 19:1; Job 38:4-33; Isaiah 40:22, 26; Hebrews 1:3; Colossians 1:17; Revelation 4:11.*

20. Does the universe reveal God's character?

Yes. Scripture teaches that the heavens declare the glory of God and that His invisible qualities are seen clearly in the things He has made. The intelligibility, relationality, rationality, and beauty of the universe reflect God's nature. A lawful universe reflects divine faithfulness. A relational universe reflects divine love. An intelligible universe reflects divine wisdom. Creation is a revelation of the Creator. *References: Psalm 19:1–4; Romans 1:20.*

SECTION III

HUMAN NATURE, SOUL, SPIRIT, AND CONSCIOUSNESS

(Questions 21–30)

21. What does it mean to be made in the image of God?

Being made in the image of God, **imago Dei**, means that human beings uniquely reflect certain attributes of their Creator. This is not about physical resemblance—God is spirit and has no body—but about capacities that mirror God's nature.

Rationality: Humans can think, reason, understand truth, and grasp abstract concepts. We can contemplate the nature of reality itself, engage in logic, and discover mathematical principles. This reflects God as the divine Logos, the rational mind behind all creation.

Relationality: Humans are designed for relationship—with God and with one another. We experience love, friendship, community, and communion. We are not meant for isolation. This reflects the Triune God, who exists eternally as Father, Son, and Holy Spirit in perfect relational communion.

Moral capacity: Humans possess conscience, the ability to discern right from wrong, and moral responsibility for their choices. We experience guilt, obligation, justice, and virtue. This reflects God's holiness and righteousness.

Creativity: Humans create art, music, literature, technology, and culture. We shape and cultivate the world around us. This reflects God as Creator, the one who brings order from chaos and beauty from formlessness.

Dominion: God gave humanity the mandate to "have dominion" over creation (*Genesis 1:26-28*). This is not tyranny but stewardship—responsible care for the world God made. We are to govern creation as God's representatives, reflecting His wise and benevolent rule.

Spiritual capacity: Humans alone among earthly creatures are oriented toward God. We worship, pray, and seek transcendence. Our spirits can commune with the Holy Spirit. We are made for relationship with the divine.

Critically, the image of God is **inherent, not earned**. It is not based on intelligence, productivity, or moral performance. Every human being—from conception to natural death, regardless of ability or capacity—bears God's image fully. This is the foundation of universal human dignity and the basis for treating every person as sacred.

The image was marred by sin but not erased. Christ, the perfect image of God (*Colossians 1:15*), restores what was broken through redemption. Believers are being renewed in knowledge and righteousness after the image of their Creator (*Colossians 3:10*). *References: Genesis 1:26-27; Psalm 8:4-6; James 3:9; Colossians 1:15; 3:10.*

22. How is the soul different from the brain?

The brain is a physical organ; the soul is the immaterial form of the person. Scripture shows that the soul animates the body and survives after death. The soul uses the brain as its instrument for thought, memory, and perception. Modern physics does not reduce consciousness to mere matter. Conscious experience cannot be explained by physical processes alone. The soul is the seat of identity, reason, and freedom and cannot be reduced to neural activity. *References: Genesis 2:7; Matthew 10:28; Ecclesiastes 12:7.*

23. Is consciousness just an emergent property of matter?

No. While the brain contributes to conscious experience, consciousness cannot be fully accounted for by physical interactions. Scripture teaches that the spirit of a person searches the depths of their being. Consciousness involves interiority, self-awareness, intentionality, and the capacity for meaning—qualities that matter alone cannot produce. Scripture and classical theology teach that consciousness arises from an immaterial dimension: the soul, which is the form of the person created by God. *References: Proverbs 20:27; 1 Corinthians 2:11.*

24. Do humans possess free will?

Yes. Scripture repeatedly affirms human freedom. While God knows all things, He does not coerce the human will. Free will is the capacity to act according to reason, desire, and moral judgment. God's creative design provides genuine space for choice. Even within physical law, human decisions operate on a higher level of causation. Freedom is essential for love, moral growth, and responsibility. *References: Deuteronomy 30:19; Joshua 24:15; John 7:17.*

25. How does the human spirit relate to the Holy Spirit?

The human spirit is the aspect of a person that is capable of communion with God. The Holy Spirit is the divine presence who enlightens, strengthens, convicts, and renews. Scripture teaches that the Spirit testifies with our spirit and intercedes within us. The human spirit is designed for relationship with the Creator, and the Holy Spirit is the divine agent who makes this relationship possible. *References: Romans 8:16; 1 Corinthians 6:19; Ezekiel 36:26–27.*

26. Can the human soul be destroyed?

Scripture teaches that the soul does not cease to exist. Physical death separates soul from body, but the soul continues in conscious existence. Jesus warns that God alone can destroy both soul and body in judgment, meaning that the soul is accountable and not naturally immortal in the pagan sense. Yet Scripture also teaches that God preserves the soul of the righteous. The soul persists because it derives its being directly from God, not from matter. *References: Matthew 10:28; Psalm 49:15; Ecclesiastes 12:7.*

27. Why do humans long for meaning?
The longing for meaning reflects our origin in the God who is "I AM." Scripture teaches that God has set eternity in the human heart. The search for purpose is not a psychological accident but a sign of transcendence. Human beings cannot satisfy this longing with material goods or social approval. Meaning comes from relational communion with God, for whom and by whom we were created. *References: Ecclesiastes 3:11; Psalm 42:1–2; John 17:3.*

28. Why do human beings experience moral conviction?

Moral conviction arises because the moral law is written on the heart. Scripture teaches that even those without Scripture have an innate sense of right and wrong. This is evidence of the divine image within every human being. Conscience is not evolutionary conditioning. It is the echo of God's justice within the human soul. Moral truths are objective because they reflect God's character. *References: Romans 2:14–15; Micah 6:8; Proverbs 3:3–4.*

29. How do emotions fit into the nature of the soul?
Emotions are expressions of the soul's perception of good and evil. Scripture shows that God Himself experiences divine affections such as compassion, love, wrath, and joy, though without imperfection. Human emotions reflect the relational nature of persons. They are not irrational states but responses that reveal what the heart values. Properly structured emotions guide moral action and deepen relationships. *References: John 11:35; Psalm 103:13; Zephaniah 3:17.*

30. Why is the human person both physical and spiritual?
Because God created humanity as a unity of body and soul. Scripture teaches that the body is good and that humans were meant to live embodied lives. The physical and spiritual dimensions complement each other. The body enables action in the world; the soul gives purpose and identity. Resurrection affirms this unity permanently. Physics supports the idea that embodiment is essential for interaction with the material universe, while the soul

anchors identity in God. *References: Genesis 2:7; Romans 8:11; 1 Corinthians 15.*

SECTION IV
EVIL, SUFFERING, AND FREEDOM

(Questions 31–40)

31. If God is good, why is there evil?

Evil exists because God created beings with genuine freedom. Scripture teaches that God made humanity upright, but they sought many ways. Evil is the misuse of freedom, not a flaw in God's design. Love, it is argued requires freedom, and freedom requires the possibility of rejection. God permits evil temporarily to accomplish greater purposes, including the development of virtue, the revelation of His justice, and the demonstration of His mercy. The cross is the ultimate example: humanity commits the greatest evil, yet God turns it into salvation. *References: Genesis 50:20; Romans 8:28; Ecclesiastes 7:29.*

32. Why does God allow suffering?

God allows suffering because He brings good from it in ways that would be impossible without it. This does not mean suffering itself is good—it is an evil consequence of living in a fallen world. But God's providence ensures that even suffering serves His redemptive purposes.

Suffering refines character. Scripture repeatedly teaches that trials produce perseverance, character, and hope (*Romans 5:3-5*). Like gold refined by fire, believers are purified through hardship. Suffering exposes what is superficial and strengthens what is genuine. It teaches patience, humility, and dependence on God.

Suffering reveals where our trust truly lies. In comfort, we can coast on self-sufficiency and forget our need for God. Suffering strips away our illusions of control and drives us to prayer. It teaches us that God alone is our refuge and strength (*Psalm 46:1*).

Suffering deepens compassion. Those who have suffered are uniquely equipped to comfort others in similar trials (*2 Corinthians 1:3-7*). Pain creates empathy. Hardship breaks down pride and opens hearts to the suffering of others. Through shared suffering, communities are formed and strengthened.

Suffering teaches us to long for what is eternal. When earthly securities fail, we learn to set our hope on heaven. Paul writes that "this light momentary affliction is preparing for us an eternal weight of glory beyond all comparison" (*2 Corinthians 4:17, ESV*). Suffering reorients our desires toward what cannot be shaken.

Suffering magnifies God's grace. In weakness, God's power is perfected (*2 Corinthians 12:9*). When we are utterly dependent, God's provision becomes undeniable. His comfort in the midst of pain demonstrates His faithfulness more powerfully than a thousand sermons.

Suffering can lead to repentance. Sometimes God allows hardship to awaken people from spiritual complacency or to call them back from destructive paths. The Prodigal Son came to himself in a pigsty (*Luke 15:17*). Difficulty can be the means God uses to save.

None of this means we should seek suffering or be passive in the face of injustice. We are called to relieve suffering where we can, to fight against evil, and to pursue justice. But when suffering comes, and it will, God's people can trust that He is present within it, working purposes that transcend our understanding.

The ultimate answer to the question of suffering is the cross. There, God Himself entered into human anguish, bore the weight of sin and death, and demonstrated that no suffering is beyond His redemptive reach. The resurrection declares that suffering does not have the final word. *References: Romans 5:3-5; 8:28; 2 Corinthians 1:3-7; 4:16-18; 12:9; James 1:2-4; 1 Peter 1:6-7; Hebrews 12:7-11.*

33: Why does God not stop natural disasters?

Natural processes such as tectonics, weather systems, and biological cycles are essential for a habitable world. The same forces that bring rain for crops can bring floods. The same plate tectonics that recycle nutrients and regulate climate can cause earthquakes. A world governed by consistent natural laws enables life, growth, and human freedom to exist at all.

Scripture teaches that creation itself has been affected by human sin. Paul writes that "the creation was subjected to futility" and that it "groans together in the pains of childbirth" awaiting redemption (*Romans 8:20-22, ESV*). Christian traditions have understood this passage in different ways:

The Eastern Orthodox tradition emphasizes the cosmic dimensions of the Fall, teaching that humanity's role as priest of creation means that when humanity fell, all creation fell with it. The natural world shares in humanity's alienation from God and will share in humanity's restoration.

The Western Catholic and Reformed traditions typically understand this passage to mean that while the natural order itself remains good, it is now experienced by fallen humanity in a context of suffering and death. The "groaning" reflects the incompleteness of creation awaiting its final redemption, not necessarily a fundamental change in natural processes themselves.

All traditions agree on several key points:

The present world is not in its final, perfected state.

Natural suffering points humanity toward our need for God and our hope in the resurrection.

God's final answer to natural evil is not the removal of the natural order but its transformation and renewal into "new heavens and a new earth" (*Revelation 21:1*).

A world without consistent natural laws would not be livable or intelligible. The stability that allows us to build, learn, and act would vanish. Natural events that sometimes bring destruction are inseparable from the same lawful structure that makes life, freedom, and moral action possible.

Scripture also shows that God uses even disasters to display His mercy, inspire compassion, awaken hearts to eternal realities, and call communities to repentance. Yet suffering from natural causes remains a consequence of living in a world not yet fully redeemed. A world groaning in anticipation of the day when God will make all things new. *References: Romans 8:18-23; Revelation 21:1-5; Job 38-41; Genesis 3:17-19; 2 Peter 3:10-13.*

34. If God knew people would sin, why create them?

Foreknowledge does not eliminate freedom. Scripture teaches that God desires relationship, love, and righteousness. Creating free beings means creating beings who can choose to disobey. God judged that a world with freedom and redemption is better than a world without freedom at all. The glory revealed in salvation, grace, and moral growth outweighs the reality of sin. *References: Ephesians 1:4–10; 2 Timothy 1:9.*

35. Can evil ever defeat God's purpose?

No. Scripture teaches that God's purposes stand forever and that no plan of His can be thwarted. Evil may cause suffering and injustice, but God ultimately overcomes it through judgment and redemption. Even acts of evil become part of the larger story that reveals God's sovereignty and goodness. The resurrection of Christ is the definitive proof that evil cannot overcome God. *References: Job 42:2; Isaiah 46:10; Acts 2:23–24.*

36. Why doesn't God destroy the devil now?

Scripture teaches that God allows spiritual evil to persist for a time for the sake of human freedom and the full revelation of justice. Evil is being permitted until its defeat demonstrates God's righteousness and teaches creation the consequences of rebellion. The devil's final destruction is promised and certain. Until that time, God limits Satan's power and uses even his schemes to accomplish greater good. *References: Revelation 20:10; Romans 16:20; Luke 22:31–32.*

37. Why do the innocent suffer?

Scripture teaches that we live in an interconnected creation where actions have consequences that extend beyond the individual. Innocent suffering is often the consequence of human sin, natural conditions, or the fallen state of the world. Scripture teaches that God is near to the brokenhearted and that He brings good from suffering, especially for the innocent who trust in Him. In the final judgment, every injustice will be overturned. *References: Psalm 34:18–19; Romans 8:28; Revelation 21:4.*

38. How can God judge people fairly?

God judges with perfect knowledge, understanding, and justice. Scripture teaches that God sees the heart, considers every circumstance, and judges righteously. Human judgment is partial; God's judgment is complete. He never punishes ignorance as rebellion nor weakness as malice. His judgment is rooted in truth and love. No one will ever be able to claim that God has been unfair. *References: 1 Samuel 16:7; Psalm 96:13; Romans 2:16.*

39: Why is hell necessary?

The doctrine of hell has been understood with varying emphases across Christian history, but all orthodox traditions affirm that it is the final state of those who reject God's grace.

The Biblical Foundation:

Scripture clearly teaches that there is eternal separation from God for those who refuse Him. Jesus speaks more about hell than anyone else in the New Testament, describing it as "outer darkness" *(Matthew 8:12)*, "eternal fire" *(Matthew 25:41)*, and a place where "their worm does not die and the fire is not quenched" *(Mark 9:48)*. Paul writes of "eternal destruction, away from the presence of the Lord" *(2 Thessalonians 1:9)*. Revelation describes the "lake of fire" and "the second death" *(Revelation 20:14-15)*.

Why Hell Exists:

All Christian traditions agree that hell exists because God is both perfectly just and perfectly respectful of human freedom. God desires all people to be saved (*1 Timothy 2:4; 2 Peter 3:9; Ezekiel 18:23*), yet He does not override human will. C.S. Lewis captured this paradox: "There are only two kinds of people in the end: those who say to God, 'Thy will be done,' and those to whom God says, in the end, 'Thy will be done.'" Hell is the tragic respect God gives to those who persistently refuse His love.

Different Traditional Understandings:

Eastern Orthodox theology often emphasizes that heaven and hell are not different locations but different experiences of the same divine presence. God is love (*1 John 4:8*), and His presence fills all reality. Those who have opened themselves to God's love experience His presence as joy and light; those who have refused Him experience that same presence as torment. The "fire" of God's love purifies those who desire Him but burns those who reject Him.

Roman Catholic teaching affirms that hell is eternal separation from God, freely chosen by the damned. The Catechism states that "the chief punishment of hell is eternal separation from God" (CCC 1035). Catholic theology has also developed the concept of varying degrees of punishment based on the gravity of one's sins and one's knowledge of God's will, reflecting God's perfect justice.

Protestant traditions generally emphasize hell as the just consequence of sin and rejection of Christ. Reformed theology stresses God's sovereignty and justice in judgment, while Arminian theology emphasizes human free will in accepting or rejecting salvation. Both agree that hell demonstrates the seriousness of sin and the necessity of Christ's atonement.

Contemporary Questions:

Some Christians in recent decades have proposed alternative views such as annihilationism (the idea that the unsaved cease to exist) or universal reconciliation (the hope that all will eventually be saved). While these views attempt to address the difficulty of eternal conscious punishment, they stand outside historic Christian orthodoxy, which has consistently affirmed the reality of eternal punishment based on Scripture's clear teaching.

The Difficulty:

No Christian should speak of hell without sorrow. It is the most sobering doctrine in Scripture.

Yet it reveals three essential truths:

Human choices have eternal significance.

God's justice is real and perfect.

The gospel is urgent.

Christ came to save us from this fate. Hell magnifies both the horror of sin and the wonder of grace.

God is not vindictive. He does not desire anyone's damnation. But He honors the freedom He gave for the sake of love. The existence of hell is the dark shadow cast by the light of human dignity and freedom. In the end, no one will be able to say that God was unfair. *References: Matthew 25:31-46; 2 Thessalonians 1:5-10; Revelation 20:11-15; Luke 16:19-31; Matthew 13:36-43; Ezekiel 18:23, 32; 2 Peter 3:9; 1 Timothy 2:4.*

40. Is God responsible for evil?

No. God is the author of being, not the author of evil. Scripture teaches that God is light and in Him there is no darkness. Evil is not a created substance; it is a lack, distortion, or privation of the good. It arises when creatures misuse the freedom God gave for love. God permits evil, but He does not cause it. His holiness remains perfect, and His purposes remain good. *References: James 1:13–17; 1 John 1:5; Deuteronomy 32:4.*

SECTION V

SCRIPTURE, THEOLOGY, AND DOCTRINE

(Questions 41–50)

41. Is the Bible historically reliable?

Yes. Scripture presents itself as historical revelation rooted in real events, real people, and real covenants. The Old and New Testaments contain genealogies, dates, locations, and eyewitness testimonies. Archaeology continues to confirm biblical names, cultures, and events. The New Testament Gospels were written within living memory of Christ's ministry, and their central claims are corroborated by early creeds and independent witnesses. Scripture is not myth but the record of God's involvement in history. *References: Luke 1:1–4; 1 Corinthians 15:3–8; Isaiah 46:9–10.*

42. How should Scripture be interpreted?

Scripture should be interpreted according to its literary context, historical background, and theological purpose. The Bible contains narrative, poetry, prophecy, letters, and wisdom literature, each requiring appropriate reading. Scripture interprets Scripture, and the meaning of a passage is illuminated by the broader biblical narrative. Ultimately, Scripture is understood through reverent study, sound reasoning, and the guidance of the Holy Spirit. *References: 2 Timothy 3:16–17; Luke 24:27; Psalm 119:105.*

43. What does it mean that Scripture is inspired?

Inspiration means that God guided human authors such that their writings communicate His word without error in all that they affirm. Scripture is fully divine and fully human, just as Christ is fully divine and fully human. The personalities, vocabularies, and contexts of the authors remain, but the final product is the trustworthy revelation of God. Inspiration is not mechanical dictation but the harmonious work of the Spirit through human minds. *References: 2 Timothy 3:16; 2 Peter 1:20–21.*

44. Is the Old Testament still relevant?

Yes. Jesus and the apostles treated the Old Testament as the authoritative word of God. The Old Testament lays the foundation for the New by revealing God's character, His covenant with Israel, His promises of a Messiah, and His moral law. The New Testament does not replace the Old but fulfills and completes it. The entire Bible reveals a single story of creation, fall, redemption, and restoration. *References: Matthew 5:17–18; Romans 15:4; Luke 24:44.*

45. Is the God of the Old Testament different from the God of the New Testament?

No. God is unchanging. Scripture reveals one God who is both just and merciful. The Old Testament contains profound expressions of God's compassion and patience, while the New Testament contains warnings of judgment. The apparent differences arise from differing historical contexts and covenantal roles, not from contradictory divine character. The same God who judged sin also provided salvation. *References: Malachi 3:6; Hebrews 13:8; Exodus 34:6–7.*

46. Why is Jesus called the Word (Logos)?

Jesus is called the Word because He is the divine rationality, wisdom, and self-expression of God. Through Him, all things were made. The Word reveals God perfectly and makes Him known. The ordered and relational structure of the universe reflects this truth: God creates through the Logos, and the Logos sustains the order and intelligibility of creation. The incarnation is the Word entering the world He created. *References: John 1:1–14; Colossians 1:15–17; Hebrews 1:1–3.*

47. How can Jesus be both God and man?

The incarnation unites the divine and human natures in one person without mixture, confusion, or division. Scripture teaches that the Word became flesh and dwelt among us. Jesus has a full human nature and a full divine nature. As God, He is eternal and uncreated. As man, He is like us in every way except sin. The incarnate Christ is the perfect mediator between God and humanity. *References: John 1:14; Philippians 2:5–11; Hebrews 4:15.*

48. Why did Jesus have to die?

Jesus died to reconcile humanity to God. Scripture teaches that sin separates us from God and that the wages of sin is death. Jesus, the sinless one, bore the consequences of humanity's rebellion, satisfying divine justice and revealing divine love. His death is not simply an example but an atonement that restores relationship with God. The resurrection vindicates His sacrifice and conquers death. *References: Isaiah 53; Romans 5:6–11; 1 Peter 2:24.*

49. What is the role of the Holy Spirit in the Christian life?

The Holy Spirit indwells believers, guiding them into truth, empowering them for service, convicting them of sin, and producing spiritual fruit. Scripture portrays the Spirit as the Comforter, Advocate, and Helper. The Spirit also illuminates Scripture, strengthens faith, and unites believers with Christ. In the relational universe, the Spirit is the divine presence who integrates human life into the life of God. *References: John 14:16–17; Romans 8:9–16; Galatians 5:22–25.*

50. What is the Church, and why does it matter?

The Church is the community of believers united to Christ and to one another through the Holy Spirit. Scripture describes the Church as the body of Christ, the temple of the Spirit, and the family of God. The Church is not merely an institution but a living organism called to worship, teach, serve, and bear witness to the truth. The Church matters because it is the visible expression of

God's presence and mission in the world. *References: Ephesians 2:19–22; Acts 2:42–47; 1 Corinthians 12:12–27.*

SECTION VI

ANGELS, DEMONS, AND THE UNSEEN REALM

(Questions 51–60)

51. Do angels really exist?

Yes. Scripture consistently teaches the existence of angels as spiritual beings created by God to serve Him and minister to His people. Angels appear throughout both Testaments in roles of protection, communication, judgment, and worship. They are rational and personal, yet immaterial. The relational structure of the universe leaves room for non-physical intelligences whose mode of existence differs from our own. *References: Hebrews 1:14; Psalm 91:11; Luke 1:26–38.*

52. What are angels made of?

Angels are immaterial spirits. Scripture describes them as winds or flames of fire, images that convey energy, purity, and mobility rather than physical composition. Angels do not possess bodies by nature, though they can appear in bodily form when God wills. Their essence is spiritual, and their power and knowledge exceed those of humans, although they remain finite and dependent on God. *References: Hebrews 1:7; Psalm 104:4.*

53. What do angels do?

Angels worship God, carry out His commands, protect His people, deliver messages, and serve as agents of divine justice. Scripture shows angels rejoicing at creation, ministering to the faithful, strengthening Jesus in Gethsemane, and announcing the resurrection. Their activity reflects the relational nature of God's governance, where spiritual beings participate in His work within creation. *References: Genesis 28:12; Luke 22:43; Matthew 28:2–7.*

54. Do humans become angels after death?

No. Scripture teaches that angels and humans are distinct orders of creation. Humans are embodied souls created in the image of God, destined for resurrection. Angels are spirits who do not marry or reproduce. After death, humans do not transform into angels. Their destiny is resurrection and eternal life in union with God. Angels and redeemed humanity coexist but remain distinct. *References: Hebrews 2:5–7; Luke 20:36; 1 Corinthians 6:3.*

55. What are demons?

Demons are fallen angels who rebelled against God. Scripture teaches that some angels chose to reject God's authority and were expelled from His presence. These beings now oppose God's purposes and seek to distort and destroy what He loves. They remain under God's ultimate authority and cannot act beyond His permission. Their rebellion reveals the seriousness of angelic freedom. *References: Revelation 12:7–9; 2 Peter 2:4; Jude 6.*

56. Why did God allow the angels to fall?

Angels were created with freedom; just as human beings were. Scripture teaches that pride was the root of angelic rebellion. God permitted this fall for reasons consistent with His justice and the greater narrative of redemption. The existence of fallen angels demonstrates the reality of spiritual freedom and serves as a warning of the consequences of rejecting God. God uses even their rebellion to reveal His righteousness and mercy. *References: Isaiah 14:12–15; Ezekiel 28:14–17.*

57. How do demons influence the world?

Demons influence the world through deception, temptation, distortion, and spiritual oppression. Scripture teaches that they blind minds, promote falsehood, sow division, and oppose the work of God. Their power is limited and must yield to the authority of Christ. Believers are instructed to resist the devil through truth, prayer, righteousness, and the power of the Holy Spirit. *References: Ephesians 6:10–18; 1 Peter 5:8–9; James 4:7.*

58. Can Christians be possessed by demons?

Scripture teaches that Christians are sealed with the Holy Spirit, and therefore demons cannot possess a believer. However, believers may experience spiritual attack, temptation, or oppression. Possession implies ownership, and Christians belong to Christ. The authority of Jesus protects His followers from domination by evil spiritual forces, though vigilance and faith are always necessary. *References: Ephesians 1:13; Romans 8:9–11; 1 John 4:4.*

59. What is spiritual warfare?

Spiritual warfare is the ongoing conflict between the forces of God and the forces of evil. Scripture teaches that believers struggle not against flesh and blood but against spiritual powers. This warfare is fought through truth, righteousness, prayer, faith, and reliance on the Holy Spirit. The battle is not waged through fear or superstition but through the strength of Christ, who has already triumphed over the powers of darkness. *References: Ephesians 6:12–18; Colossians 2:15.*

60. What is the ultimate destiny of angels and demons?

Faithful angels will continue to serve God in the renewed creation, participating in worship and divine governance. Scripture portrays them rejoicing in God's eternal kingdom. Demons will face final judgment and be cast into the lake of fire, no longer able to oppose God or afflict His people. The ultimate destiny of all spiritual beings reflects God's justice and holiness. *References: Matthew 25:41; Revelation 20:10; Hebrews 12:22–24.*

SECTION VII

AFTERLIFE, RESURRECTION, AND ETERNITY

(Questions 61–70)

61. What happens the moment a person dies?

Scripture teaches that when the body dies, the soul separates from the body and enters a state of conscious existence. For believers, this means immediate presence with Christ. For those who reject God, it means separation from His presence. Physical death is not the end of personal existence because the soul is sustained directly by God. The body returns to dust, but the soul continues in awareness until the resurrection. *References: 2 Corinthians 5:8; Luke 23:43; Ecclesiastes 12:7.*

62. What is the intermediate state?

The intermediate state is the condition of the soul between physical death and the resurrection of the body. Scripture describes believers as being with Christ yet awaiting the fullness of resurrection. It is a conscious state of rest and anticipation. Those who reject God experience conscious separation from Him. The intermediate state is temporary, pointing toward the final renewal of creation. *References: Philippians 1:23; Revelation 6:9–11; Hebrews 12:23.*

63. Will everyone be resurrected?

Yes. Scripture teaches that all people, both the righteous and the unrighteous, will be raised bodily. Resurrection is universal, but the outcomes differ. For believers, resurrection results in eternal life in union with God. For those who reject Him, resurrection leads to judgment. The resurrection demonstrates the permanence of human embodiment, and the value God places on the human person. *References: John 5:28–29; Daniel 12:2; Acts 24:15.*

64. What will the resurrection body be like?

The resurrection body will be physical, immortal, incorruptible, and glorified. Scripture teaches that believers will bear the likeness of Christ's resurrected body. It will not be limited by decay, disease, or death. The resurrection body will retain personal identity while being transformed. It will be suited for life in the new creation, where righteousness dwells and God's presence is fully experienced. *References: 1 Corinthians 15:42–49; Philippians 3:20–21; Luke 24:36–43.*

65. What is heaven?

Heaven is the dwelling place of God, the realm in which God's will is perfectly done, and the future home of the redeemed. It is not merely a spiritual dimension but the fullness of God's kingdom in a transformed creation. Scripture describes heaven as a place of joy, worship, peace, and unbroken fellowship with God. Heaven is defined by the presence of God rather than by ethereal imagery. *References: Revelation 21:1–4; John 14:1–3; Psalm 16:11.*

66. What is hell?

Hell is the state of eternal separation from God. Scripture teaches that hell is a place of conscious judgment for those who reject God's love and choose rebellion. It is not arbitrary but the natural consequence of refusing the source of goodness, truth, and life. Hell is the final confirmation of human freedom. God does not desire anyone to perish, but He honors the choices of those who reject Him. *References: Matthew 25:46; 2 Thessalonians 1:9; Ezekiel 18:23.*

67. Will the earth be destroyed or renewed?

Scripture teaches that the earth will be renewed, not annihilated. The language of fire in prophecy refers to purification and transformation, not obliteration. God intends to restore creation to its original beauty and perfection. The new heavens and the new earth will be free from corruption, death, and decay. The renewed world will unite heaven and earth under God's eternal rule. *References: Romans 8:19–23; 2 Peter 3:10–13; Revelation 21:1.*

68. What will we do in eternity?

Scripture teaches that eternity will involve worship, meaningful work, joyful relationships, learning, creativity, and communion with God. The redeemed will reign with Christ and participate in the flourishing of the renewed creation. Eternity is not static or monotonous. It is the fullness of life for which humanity was created. Every gift and capacity will be perfected and used without distortion. *References: Revelation 22:3–5; Matthew 25:23; Isaiah 65:17–25.*

69. Will we recognize one another in eternity?

Yes. Scripture indicates that personal identity continues after death and resurrection. Jesus' disciples recognized Him after His resurrection. Moses and Elijah retained their identities at the Transfiguration. Believers will know and be known. Resurrection does not erase individuality but restores it. Relationships will be purified, healed, and deepened in the presence of God. *References: 1 Corinthians 13:12; Matthew 17:1–3; Luke 24:31.*

70. Does eternal life begin after death or now?

Eternal life begins now. Scripture teaches that eternal life is knowing God and Jesus Christ whom He has sent. This relationship begins in this life and continues into eternity. While the fullness of eternal life awaits resurrection, the life of the age to come is experienced in the present through union with Christ. Believers already taste the powers of the coming age. *References: John 17:3; 1 John 5:11–13; Ephesians 2:4–7.*

SECTION VIII
MEANING, PURPOSE, AND THE MORAL LIFE

(Questions 71–80)

71. What is the purpose of human life?

Scripture teaches that the purpose of human life is to know God, love Him, reflect His character, and share in His work within creation. Humans are made in the image of God, which means they are relational, rational, moral, and capable of love. We are participants in God's ongoing work in creation, mirrors of God's creative Word. Meaning comes from living in relationship with the One who is the source of all being. Our life has meaning not because of what we accomplish, but because we bear the image of the One whose being is love and truth. The purpose of life is to grow into that likeness: to know the truth, to love the good, and to live in a way that reflects the One who created us. *References: Ecclesiastes 12:13; Micah 6:8; John 17:3.*

72. Is human life inherently meaningful?

Yes. Human life is inherently meaningful because it is grounded in God's creative purpose and sustained by His will. Meaning is not something we invent or construct—it is something **given** to us by the One who made us.

Meaning flows from being made in God's image. Every human being bears the **imago Dei** (*Genesis 1:26-27*), which establishes our fundamental dignity and worth. We are not accidents of blind evolutionary processes but intentional creations of a personal God. Our existence has purpose because we were designed for relationship with the One who is Himself the source of all being and goodness.

Meaning is not dependent on circumstances. Even in suffering, limitation, poverty, or disability, human life retains its full meaning and dignity. A person's worth is not measured by productivity, intelligence, social status, or physical capacity. Scripture teaches that God knows each person intimately—He knits us together in the womb (*Psalm 139:13-16*), numbers the hairs on our heads (*Matthew 10:30*), and values us more than the whole created order (*Matthew 10:31*).

Meaning is found in knowing God. Jesus defines eternal life as knowing the Father and the Son (*John 17:3*). This is not merely intellectual knowledge but intimate, relational communion. We are made to know and be known by God. When we live in relationship with Him, our existence aligns with its intended purpose.

Meaning is discovered through love. The greatest commandment is to love God with all our heart and to love our neighbor as ourselves (*Matthew 22:37-40*). In loving others, we participate in the very nature of God, who is love (*1 John 4:8*). Acts of sacrificial love, mercy, justice, and compassion are inherently meaningful because they reflect God's character.

Meaning extends beyond this life. Because human beings possess souls that endure beyond physical death, our lives have eternal significance. The choices we make, the love we give, and the truth we pursue shape us not just for this age but for the age to come. This gives even the smallest acts of faithfulness profound importance.

Meaning is not negated by death. Materialists struggle to find meaning in a universe destined for heat death, where all human achievements will eventually be erased. But Scripture teaches that

God will renew creation, raise the dead, and establish His kingdom in a new heavens and new earth where righteousness dwells (*Revelation 21:1-5*). Nothing done in faith is ever wasted.

Even when we cannot see the purpose of our suffering or understand God's plan, we can trust that our lives matter to the God who created us, redeemed us, and promises to complete the good work He began in us (*Philippians 1:6*). Human life is meaningful because we exist in relationship to the God who is "I AM"—the eternal, unchanging, all-sufficient ground of all reality. *References: Genesis 1:26-27; Psalm 139:13-16; Ecclesiastes 3:11; Matthew 10:29-31; John 17:3; Philippians 1:6; Revelation 21:1-5.*

73. What is the relationship between God's nature and human purpose?

Scripture teaches that human purpose flows directly from God's nature as Creator and Sustainer. Because God is personal, relational, and loving, He creates persons who can know Him and respond to Him. Because He is rational (the Logos), He creates rational beings who can understand truth. Because He is good, He creates beings capable of recognizing and choosing goodness.

Purpose is not imposed externally but flows from our created nature. God calls creation into being, and human beings respond within that relationship. We are made to reflect God's character, to participate in His ongoing work, and to grow toward the perfection for which we were designed. Our purpose aligns with reality itself because reality is grounded in the God who created us for relationship with Himself. *References: Acts 17:26-28; Ephesians 2:10; Colossians 1:16-17; Hebrews 11:3.*

74. What is sin?

Sin is the rejection of God's will, the distortion of His good creation, and the attempt to ground one's identity apart from God. Sin is not merely rule-breaking but a rupture in relationship. It blinds the mind, disorders desire, and sabotages flourishing. Scripture views sin as both a power and a personal act. Sin prevents us from reflecting God's image as intended. *References: Romans 3:23; Isaiah 59:2; 1 John 3:4.*

75. Why is sin such a serious matter?

Sin is serious because it separates humanity from God, undermines human flourishing, corrupts relationships, and damages creation. Sin contradicts the nature of the God who is truth, goodness, and life. Sin is the choice to break communion with the One in whom we live and move and have our being (*Acts 17:28*). Sin is not just wrong; it is destructive to our relationship with God and to the fabric of creation itself. *References: Romans 6:23; James 1:14–15; Galatians 6:7–8.*

76. What is repentance?

Repentance is the turning of the heart and mind toward God. It involves acknowledging sin, grieving its effects, and choosing to walk in a new direction empowered by grace. Repentance is not merely remorse; it is transformative reorientation. Scripture calls repentance a gift from God and the beginning of renewed relationship. It restores harmony between the human person and God's will. *References: Acts 3:19; 2 Corinthians 7:10; Ezekiel 18:30–32.*

77. Can people truly change?

Yes. Through the transforming work of the Holy Spirit, people can be renewed in mind, heart, and character. Scripture teaches that those who are in Christ are new creations, with old patterns passing away. Transformation is a lifelong process but a real one. Change is possible not through human willpower alone but through divine assistance and the renewing power of truth. *References: 2 Corinthians 5:17; Romans 12:1–2; Ezekiel 36:26–27.*

78. Why is moral obedience important?

Moral obedience aligns human life with God's character and design. Obedience is not a burden but a path to flourishing. Scripture teaches that God's commandments are expressions of His love and wisdom. When we obey, we participate in the harmony and order of God's creation. Obedience reflects relational alignment with the moral structure God has woven into reality—it is living in accordance with how we were designed to function *References: John 14:15; Psalm 119:1–11; Deuteronomy 10:12–13.*

79. What is the role of conscience?

Conscience is the inner moral sense God has placed within every human being. It can be shaped by truth or distorted by sin, but in its proper form it bears witness to the divine moral law. Conscience urges what is right and warns against what is wrong. Scripture teaches that conscience must be informed by God's Word and kept sensitive through humility and obedience. *References: Romans 2:14–15; 1 Timothy 1:5; Hebrews 10:22.*

80. How does love fulfill the moral law?

Love fulfills the moral law because love embodies God's character. Jesus taught that all the commandments are summed up in loving God with all the heart and loving one's neighbor as oneself. Love is not sentiment but active, self-giving alignment with God's goodness. When love governs the heart, righteousness naturally flows from it. Love is the highest expression of moral coherence because God Himself is love. *References: Matthew 22:36–40; Romans 13:8–10; 1 John 4:7–12.*

SECTION IX

SCIENCE, SUFFERING, AND DIVINE PROVIDENCE

(Questions 81–90)

81: Why does suffering exist?

Suffering entered the world through human sin and the resulting disruption of creation's original harmony. Scripture teaches that God created the world "very good" (*Genesis 1:31*), but that goodness has been marred by the Fall.

The Origin of Suffering:

When humanity rebelled against God, the consequences rippled through all creation. God told Adam, "Cursed is the ground because of you; in pain you shall eat of it all the days of your life" (*Genesis 3:17, ESV*). Paul explains that "the creation was subjected to futility" and "has been groaning together in the pains of childbirth until now" (*Romans 8:20-22, ESV*).

Christian traditions understand this cosmic effect of sin differently:

Eastern Orthodox theology emphasizes humanity's priestly role in creation. Humanity was meant to mediate God's grace to the physical world. When humanity fell, all creation fell with it because the priest abandoned his post. The natural world shares in both humanity's fallenness and, through Christ, humanity's redemption. Suffering is thus deeply connected to the broken relationship between humanity and creation.

Western (Catholic/Protestant) theology
tends to emphasize that God's curse involved subjecting creation to "frustration" or "futility"—meaning that natural processes, while still good in themselves, now operate in a context where they bring pain to fallen humanity. Death, thorns, toil, and natural hardship entered human experience even though the laws of nature remained essentially the same. The change is in the relationship between humanity and the natural world.

Redemptive Suffering:
While suffering originated in sin, Scripture teaches that God uses it redemptively.

Suffering can:

Refine character and produce perseverance.
Romans 5:3-5; James 1:2-4)

Deepen dependence on God.
(2 Corinthians 12:9-10)

Awaken compassion and empathy.
(2 Corinthians 1:3-7)

Reveal what we truly value.
(1 Peter 1:6-7)

Draw us toward eternal realities.
(2 Corinthians 4:16-18)

The consistent testimony of Scripture and Christian experience is that God can bring good from suffering even when He did not cause the suffering itself (*Romans 8:28*). This is not to say suffering is good, rather it remains an evil consequence of living in a fallen world. However, it is only God's redemptive power that can transform even evil into an instrument of grace.

The Physical Laws and Suffering:
The world operates according to stable, consistent natural laws. These laws make life, learning, and meaningful action possible. But they also mean that certain actions have painful consequences: fire

burns, gravity pulls, diseases spread. A world without these regularities would be unlivable and incomprehensible. We could not plan, build, or act if the rules constantly changed. The same physical consistency that enables human flourishing also allows for physical suffering.

Three Essential Truths:

- Suffering was not part of God's original design but entered through sin.
- God does not abandon us in suffering; Christ entered into it Himself (Hebrews 4:15).
- Suffering will not last forever; God promises ultimate restoration (Revelation 21:4).

God permits suffering for a season but has already set in motion its final defeat through Christ's death and resurrection. The cross demonstrates that God does not watch suffering from a distance—He entered into it, bore it, and conquered it. Our present groaning awaits the fulfillment of that victory. *References: Genesis 3:17 19; Romans 8:18-28; 2 Corinthians 4:16-18; James 1:2-4; 1 Peter 1:6-7; Romans 5:3-5; Hebrews 4:15; Revelation 21:4; Job 1-2; Psalm 73.*

82. Why do good people suffer?

The suffering of the righteous is one of the most painful mysteries in human experience. Yet Scripture addresses this directly, both through teaching and through the stories of God's faithful people who endured great hardship.

Suffering Is Not Always Punishment for Sin

Jesus explicitly rejected the idea that suffering is always tied to personal wrongdoing. When His disciples asked about a man born blind, "Who sinned, this man or his parents?" Jesus answered, "It was not that this man sinned, or his parents, but that the works of God might be displayed in him" *(John 9:2-3, ESV)*. God's purposes in allowing suffering extend far beyond simple cause-and-effect punishment.

Job is the clearest biblical example. He was "blameless and

upright, one who feared God and turned away from evil" *(Job 1:1, ESV)*, yet he lost his children, his wealth, and his health. His friends insisted he must have sinned, but God vindicated Job and rebuked those who claimed to know why he suffered. The book of Job teaches that God's wisdom exceeds our understanding and that suffering cannot be reduced to a simple moral calculus.

The Righteous Have Always Suffered

The Bible is filled with examples of godly people who endured tremendous hardship:

Joseph was betrayed by his brothers, sold into slavery, falsely accused, and imprisoned—yet he remained faithful, and God used his suffering to save nations *(Genesis 37-50)*.

David was anointed king but spent years fleeing for his life from Saul, even though he had done nothing wrong *(1 Samuel 19-31)*.

The prophets were rejected, persecuted, and often killed for speaking God's truth *(Hebrews 11:36-38)*.

The apostles were beaten, imprisoned, and martyred for proclaiming the gospel *(Acts 5:40-41; 2 Corinthians 11:23-28)*.

Jesus Himself—the only perfectly righteous person who ever lived—suffered betrayal, mockery, torture, and crucifixion. If the sinless Son of God suffered unjustly, His followers should not expect exemption from suffering.

God Is Near to the Brokenhearted

Scripture does not offer glib explanations for undeserved suffering, but it does offer profound comfort: **God is present with those who suffer.**

"The LORD is near to the brokenhearted and saves the crushed in spirit" *(Psalm 34:18, ESV)*. God does not abandon His people in their pain. He sees, He knows, and He cares deeply. The righteous who endure trials can be confident that God is with them in the fire *(Isaiah 43:2)*, walking with them through the valley of the shadow of death *(Psalm 23:4)*.

Suffering Produces Perseverance and Hope

While this does not make suffering pleasant, Scripture teaches that God works through it. "We rejoice in our sufferings, knowing that suffering produces endurance, and endurance produces character, and character produces hope" (*Romans 5:3-4, ESV*). The faithful who endure trials with trust in God are refined, strengthened, and drawn closer to Him.

Peter writes to persecuted believers: "After you have suffered a little while, the God of all grace, who has called you to his eternal glory in Christ, will himself restore, confirm, strengthen, and establish you" (*1 Peter 5:10, ESV*). Suffering is temporary. God's restoration is eternal.

A Crown of Life Awaits

Jesus promises, "Blessed are you when others revile you and persecute you and utter all kinds of evil against you falsely on my account. Rejoice and be glad, for your reward is great in heaven" (*Matthew 5:11-12, ESV*). Those who suffer unjustly for righteousness' sake will be vindicated. Every tear will be wiped away. Every injustice will be made right. God will reward those who remained faithful through the fire.

The Final Justice

In the end, no one will be able to say God was unfair. He sees every act of cruelty. He knows every hidden injustice. He will bring every deed into judgment (*Ecclesiastes 12:14*). Those who suffer innocently can trust that God is the righteous judge who will set all things right. The wicked will face judgment. The faithful will receive their inheritance.

The suffering of the righteous does not indicate God's absence or indifference. It reveals that we live in a fallen world awaiting redemption, and it invites us to trust the God who suffered for us and promises to bring us through suffering into eternal joy. *References: Job 1:1; Psalm 34:18; Matthew 5:10-12; John 9:1-3; Romans 5:3-5; 1 Peter 5:10; Hebrews 11:36-38.*

83. Does God cause suffering?

God is never the author of evil, but He allows suffering within the boundaries of His sovereign wisdom. Scripture shows that God can permit trials for discipline, growth, protection, or the fulfillment of a larger purpose. His permission is never arbitrary. He brings good from evil, light from darkness, and redemption from pain. God's sovereignty does not imply causation of moral evil, but His providence ensures that evil will not have the final word. *References: James 1:13–17; Hebrews 12:5–11; Genesis 50:20.*

84. How does the (IAM) approach suffering?

In the (IAM) framework, suffering emerges in a universe where freedom, relationality, and genuine agency are real. A world capable of love must also permit the possibility of harm. I AM does not learn through suffering, However humans, as agents within the relational universe, come to know both the depth of good and the reality of brokenness. This understanding highlights the dignity and weight of human freedom. *References: Deuteronomy 30:19; Romans 5:3–5; John 16:33.*

85. Can science explain suffering?

Science can describe the physical, biological, and psychological mechanisms behind suffering, but it cannot answer the moral or existential questions of why suffering matters or how it should be interpreted. Science can identify the causes of disease but not the meaning of illness. Scripture provides the moral context, grounding suffering in the realities of fallenness, love, hope, and redemption. *References: Ecclesiastes 7:13–14; Romans 8:22–28.*

86. Where is God when we suffer?

Scripture consistently teaches that God is near to the suffering, present in their pain, and active in their deliverance. In Christ, God entered into human suffering directly. He does not remain distant from human anguish. The cross reveals a God who suffers with and for His people. The Spirit comforts, strengthens, and sustains believers in their trials. God is never absent; He is most present in weakness. *References: Psalm 23:4; Isaiah 43:1–3; Hebrews 4:15.*

87. Does prayer change things?

Yes. Prayer changes things because God has chosen to work through the prayers of His people. Prayer is not merely psychological comfort or spiritual discipline—it is a real means by which God accomplishes His purposes in the world.

Prayer Is Participation in God's Will

God is sovereign, but He does not work unilaterally in every circumstance. He invites His people to participate in His work through prayer. James writes, "You do not have, because you do not ask" (James 4:2, ESV), indicating that some blessings are conditional on prayer. Jesus commands, "Ask, and it will be given to you; seek, and you will find; knock, and it will be opened to you" (Matthew 7:7, ESV). These are not empty promises—they reveal that prayer is instrumental in God's economy.

God ordains not only ends but also means. He has determined that certain outcomes will come about **through** the prayers of His people, not apart from them. Prayer is the divinely appointed means by which we partner with God in His redemptive work. When we pray, we are not informing God of something He doesn't know or convincing Him to change His mind. We are aligning ourselves with His will and participating in its fulfillment.

Scripture Demonstrates Prayer's Effectiveness

The Bible records countless instances where prayer changed circumstances:

Moses interceded for Israel, and God relented from the disaster He had planned (Exodus 32:11-14).

Hezekiah prayed when facing death, and God added fifteen years to his life (2 Kings 20:1-6).

Daniel prayed, and the course of nations was altered (Daniel 9:20-23).

The early church prayed for Peter, and he was miraculously released from prison (Acts 12:5-17).

Elijah prayed, and rain ceased for three years; he prayed again, and rain returned (James 5:17-18).

These are not anomalies but examples of the pattern Scripture establishes: "The prayer of a righteous person has great power as it is working" (James 5:16, ESV).

Prayer Changes the One Who Prays

While prayer affects external circumstances, it also transforms the one praying. In prayer, we draw near to God, and He draws near to us (James 4:8). Prayer aligns our desires with God's will, shapes our character, deepens our trust, and increases our dependence on Him. Through prayer, we become more like Christ.

Prayer teaches us humility; we acknowledge our need and God's sufficiency. It teaches us gratitude; we recognize His provision and faithfulness. It teaches us patience; we wait for His timing. Prayer is the means by which God forms us into the people He has called us to be.

Prayer Is Relational, Not Mechanical

Prayer is not a technique for manipulating God or a formula for getting what we want. It is relationship. We come to our Father knowing that He hears us, cares for us, and delights in giving good gifts to His children (**Matthew 7:11**). Sometimes the answer is "yes," sometimes "no," sometimes "wait." But we can be confident that God always answers in accordance with His perfect wisdom and love.

Paul prayed three times for the removal of his "thorn in the flesh," and God answered, "My grace is sufficient for you" *(2 Corinthians 12:9, ESV)*. The answer was not what Paul wanted, but it was what he needed. Prayer brings us into submission to God's will, trusting that His plans are better than ours.

Prayer and Providence

How can prayer change things if God is sovereign and His plans are certain? Because God's sovereignty includes the means as well as the ends. He has ordained that some things will happen **through** prayer. From the standpoint of eternity, God knows every prayer that will be prayed and has woven them into His eternal plan. From our standpoint in time, we pray, and things happen as a result.

This is not a contradiction but a mystery of God's relational providence. He invites us to participate in His work, and our participation matters. Prayer is not pointless because God is sovereign; prayer is powerful **because** God is sovereign and has chosen to work through it.

The Invitation to Pray

God commands us to pray *(1 Thessalonians 5:17; Philippians 4:6)*. He invites us to bring every concern, every need, every joy, and every sorrow before Him. Prayer changes things because the God who hears prayer is faithful, powerful, and loving. When we pray according to His will, He hears us, and we can be confident that He will grant our requests (*1 John 5:14-15*).

So pray. Pray with faith. Pray with persistence. Pray knowing that your prayers matter, that they are heard, and that they accomplish God's purposes in ways beyond your understanding. *References: Matthew 7:7-11; James 4:2; 5:16-18; Philippians 4:6; 1 Thessalonians 5:17; 1 John 5:14-15; 2 Corinthians 12:9.*

88. How does divine sovereignty work with human freedom?

Scripture presents both divine sovereignty and human freedom as true, without fully explaining how they coexist. This is not a contradiction but a mystery, one that finite minds encounter when grappling with the infinite God.

God Is Fully Sovereign

Scripture declares God's absolute sovereignty in the clearest terms:

"The LORD has established his throne in the heavens, and his kingdom rules over all" (*Psalm 103:19, ESV*).

"He does according to his will among the host of heaven and among the inhabitants of the earth; and none can stay his hand or say to him, 'What have you done?'" (*Daniel 4:35, ESV*).

"In him we have obtained an inheritance, having been predestinated according to the purpose of him who works all things according to the counsel of his will" (*Ephesians 1:11, ESV*).

God's purposes cannot be thwarted. No plan of His can be frustrated (*Job 42:2*). He works all things, not just some things, according to His will. His sovereignty is total, extending over nature, history, nations, and individual lives.

Humans Are Genuinely Free

Yet Scripture equally affirms human freedom and responsibility:

"I have set before you life and death, blessing and curse. Therefore choose life" (*Deuteronomy 30:19, ESV*).

"If it is evil in your eyes to serve the LORD, choose this day whom you will serve" (*Joshua 24:15, ESV*).

"Whoever believes in him is not condemned, but whoever does not believe is condemned already" (*John 3:18, ESV*).

Human choices are real. We are not puppets. Our decisions have consequences. We are held morally responsible for what we choose. Scripture repeatedly calls people to repent, believe, obey, and choose—commands that would be meaningless if we were not free.

The Paradox Held in Tension

How can both be true? Scripture does not offer a philosophical resolution. Instead, it holds both truths in tension:

Joseph's brothers freely chose to sell him into slavery, yet Joseph later declares, "You meant evil against me, but God meant it for good" (*Genesis 50:20, ESV*). Their sin was real; God's sovereign plan was also real.

At the cross, wicked men freely chose to crucify Jesus, yet Peter proclaims that He was "delivered up according to the definite plan and foreknowledge of God" (*Acts 2:23, ESV*). Human evil and divine purpose coexisted.

In salvation, God chose believers before the foundation of the world (*Ephesians 1:4*), yet Jesus invites, "Whoever comes to me I will never cast out" (*John 6:37, ESV*). Election and invitation are both true.

Theological Perspectives

Different Christian traditions have emphasized different aspects of this mystery:

Reformed (Calvinist) theology emphasizes God's sovereignty, teaching that He predestines all things, including who will be saved. Human choices are real but operate within the framework of God's eternal decree. God's grace is irresistible, and those He chooses will certainly come to faith.

Arminian theology emphasizes human freedom, teaching that God's sovereignty includes granting genuine libertarian free will. God knows all things, including future free choices, but does not causally determine them. His grace is resistible, and salvation depends on the free response of faith.

Molinist theology proposes that God knows not only what will happen but what **would** happen under any possible circumstance (middle knowledge). He sovereignly arranges circumstances such that people freely choose in accordance with His will.

Catholic and Orthodox theology speak of divine sovereignty and human cooperation (synergy), affirming that grace and freedom work together without compromising either.

The Mystery Remains

No theological system fully resolves the paradox. The finite mind cannot comprehend how the infinite God relates to time, causality, and freedom. What we can affirm is this:

God is sovereign. His will is supreme, His plans are certain, and His purposes will be accomplished.

Humans are free. Our choices are real, our responsibility is genuine, and our actions have consequences

God's sovereignty establishes the framework in which freedom operates. He does not violate our wills, nor does our freedom compromise His control

Both truths serve pastoral purposes. God's sovereignty comforts us—nothing is out of His control, and He works all things for the good of those who love Him (*Romans 8:28*). Human freedom challenges us—we are accountable for our choices and must actively pursue righteousness.

Practical Application

We should pray as if everything depends on God (because it does) and act as if everything depends on us (because our actions matter). We trust God's sovereign grace while taking full responsibility for our choices. We rest in His control while laboring diligently in His service.

The mystery of sovereignty and freedom invites humility. We do not need to resolve what God has not explained. We need only to trust that He is good, wise, and just—and that both His sovereignty and our freedom serve His perfect purposes. *References: Psalm*

103:19; Proverbs 16:9; Daniel 4:35; Acts 2:23; Romans 8:28-30; Ephesians 1:11; Philippians 2:12-13; Job 42:2.

89. If God controls everything, why should we act?

Because Scripture teaches that human action is meaningful. God works through human choices, not instead of them. Obedience, prayer, kindness, courage, and justice are instruments of God's will. The fact that God is sovereign gives our actions weight, not futility. We act because we are participants in God's unfolding purpose, not spectators of a predetermined script. *References: James 1:22; Micah 6:8; Esther 4:14.*

90. What hope does Christianity offer in suffering?

Christianity offers the hope of redemption, resurrection, and restoration. Suffering will not last. Scripture promises that God will wipe away every tear, death will be defeated, and creation will be renewed. The cross shows that God brings victory from apparent defeat. The resurrection guarantees that suffering does not have the final word. Hope rests not in circumstances but in the character of God. *References: Revelation 21:4; Romans 8:35–39; John 14:1–3.*

SECTION X

SALVATION, REDEMPTION, AND THE CHRISTIAN LIFE

(Questions 91–100)

91. What is salvation?

Salvation is God's work of rescuing humanity from sin, reconciling us to Himself, restoring His image in us, and bringing us into eternal life through Jesus Christ. Scripture teaches that salvation is by grace, through faith, rooted in Christ's death and resurrection. Salvation is both an event and a process: we are justified, we are being sanctified, and we will be glorified. Salvation restores the relationship broken by sin. References: *Ephesians 2:8–9; Romans 5:1–2; Titus 3:4–7.*

92. What does it mean to be "born again"?

To be born again means to receive new spiritual life through the work of the Holy Spirit. This new birth transforms the heart, renews the mind, and begins a life of growth toward Christlikeness. Jesus taught that without this rebirth, no one can enter the kingdom of God. The new birth is God's creative act within the human person, awakening faith and reorienting desire toward truth and goodness.

References: *John 3:3–8; 1 Peter 1:3; Ezekiel 36:26–27.*

93. What is justification?

Justification is God's declaration that a person is righteous in His sight through faith in Christ. It is not earned by works but received by trusting in Christ's atoning sacrifice. God imputes Christ's righteousness to the believer, forgiving sin and restoring right standing with Him. Justification is a once-for-all act grounded in the finished work of Christ. *References: Romans 3:23–26; Galatians 2:16; 2 Corinthians 5:21.*

94. What is sanctification?

Sanctification is the ongoing process by which the Holy Spirit transforms believers into the likeness of Christ. It involves the renewal of the mind, the growth of virtue, and the daily practice of obedience. Sanctification is synergistic: God works in us, and we respond with faith and effort. It is both inevitable and intentional, the outworking of a new nature created by God. *References: Philippians 2:12–13; 1 Thessalonians 4:3; Romans 12:1–2.*

95. What is glorification?

Glorification is the final stage of salvation when believers are resurrected, perfected, and fully conformed to the image of Christ. Sin, suffering, and death will no longer exist. The resurrection body will be incorruptible, immortal, and radiant with God's glory. Glorification completes God's redemptive plan for humanity and ushers believers into the eternal joy of the new creation. *References: Romans 8:29–30; 1 Corinthians 15:51–54; 1 John 3:2.*

96: Can a person lose their salvation?

This question has been answered differently across Christian traditions, and Scripture contains passages that have been interpreted to support different conclusions. Rather than pretending there is only one view, let us examine what Scripture teaches and how faithful Christians have understood it.

Points of Universal Agreement:

All orthodox Christian traditions agree on these essential truths:

- Salvation is by grace alone, through faith in Jesus Christ (Ephesians 2:8-9).
- God is faithful and His promises are sure (*2 Timothy 2:13*).
- Believers are called to persevere in faith (*Hebrews 3:14; Colossians 1:23*).
- Those who persist in deliberate, unrepentant sin show that saving faith may never have been genuine (*1 John 2:19*).

The Reformed (Calvinist) Position: Perseverance of the Saints

This tradition, drawing from Augustine and developed systematically by Calvin, teaches that those whom God has truly saved will persevere to the end. True believers may stumble and fall into sin, but they will not totally and finally fall away. Key biblical support includes:

- Jesus' promise: "I give them eternal life, and they will never perish, and no one will snatch them out of my hand" (*John 10:28, ESV*).
- Paul's confidence: "He who began a good work in you will bring it to completion" (*Philippians 1:6, ESV*).
- The sealing of the Spirit: believers are "sealed for the day of redemption" (*Ephesians 4:30, ESV*).

In this view, warnings about falling away (*Hebrews 6:4-6; 10:26-31*) serve to expose false professors and to keep genuine believers vigilant. Those who ultimately abandon faith were never truly saved to begin with (*1 John 2:19*).

The Arminian Position: Conditional Security

This tradition, articulated clearly by Arminius and Wesley, teaches that while salvation is secure in Christ, believers can through persistent unbelief and sin forfeit their salvation. Faith must be maintained.

Key biblical support includes:

Warnings to believers: "Take care, brothers, lest there be in any of you an evil, unbelieving heart, leading you to fall away from the living God" (*Hebrews 3:12, ESV*).

Paul's concern for himself: "I discipline my body and keep it under control, lest after preaching to others I myself should be disqualified" (*1 Corinthians 9:27, ESV*).

The possibility of shipwrecked faith: some have "made shipwreck of their faith" (*1 Timothy 1:19, ESV*).

In this view, God's keeping power is real, but it operates through the believer's continued faith. Perseverance is both God's work and the believer's responsibility.

The Catholic Position: Cooperation with Grace

Catholic theology teaches that salvation is a process involving both God's grace and human cooperation. Justification comes by grace through faith, but that faith must be "working through love" (*Galatians 5:6*). Mortal sin can sever the relationship established in baptism, requiring confession and restoration. Final salvation (what Catholics call "glorification") is assured only for those who persevere to the end in faith and charity. The Church's teaching emphasizes both God's sustaining grace and the necessity of the believer's ongoing cooperation with that grace.

The Eastern Orthodox Position: Synergy and Theosis

Orthodox theology speaks of salvation as synergy—cooperation between divine grace and human freedom. Salvation is understood as an ongoing process of theosis (becoming partakers of the divine nature, *2 Peter 1:4*) that continues throughout life. One can fall away from this process through persistent sin and rejection of God's grace, but one can also be restored through repentance. The emphasis is less on a legal transaction and more on a relational reality that must be maintained.

Pastoral Wisdom:

Both positions contain pastoral dangers:

- Over-emphasizing security can lead to presumption and moral laxity.

- Over-emphasizing the possibility of loss can lead to anxiety and works-righteousness.

Scripture's paradox: God keeps us, yet we must persevere—serves both to comfort the struggling believer and to warn the presumptuous. The best pastoral counsel combines both truths:

To the fearful believer struggling with sin. God is faithful. He will not let you go. Your faith, however weak, is held by His strong hand (*John 10:27-29; Jude 24-25*).

To the complacent believer presuming on grace: Take care. Continue in the faith. Work out your salvation with fear and trembling, knowing it is God who works in you (*Philippians 2:12-13; Hebrews 3:12-14*).

The Final Word:

Whatever our theological tradition, we can affirm that salvation is God's work from beginning to end, that true faith will be evidenced by perseverance, and that no one who genuinely desires to remain in Christ will be cast out (*John 6:37*). The security we have is not in our grip on God but in God's grip on us, yet that grip is experienced through our ongoing trust in Him. *References: John 10:27-29; Philippians 1:6; Ephesians 4:30; Hebrews 3:12-14; 6:4-8; 10:26-31; 1 Corinthians 9:27; 1 Timothy 1:19; 1 John 2:19; Jude 24-25; Philippians 2:12-13; Colossians 1:21-23; 2 Peter 1:10.*

97. What is the role of good works?

Good works are the fruit of salvation, not the cause of it. They flow from a transformed heart and demonstrate the reality of faith. Scripture teaches that believers are created in Christ for good works that God prepared beforehand. Good works express love, reflect God's character, and participate in His mission. They do not earn salvation but reveal the life of God within the believer. *References: Ephesians 2:10; James 2:14–18; Matthew 5:16.*

98. How should Christians live in a fallen world?

Christians are called to live with holiness, courage, humility, and wisdom. They are to resist evil, love their neighbors, and bear witness to the truth. Scripture teaches that believers are lights in the world, ambassadors for Christ, and citizens of a coming kingdom. The Christian life is a discipline of trust, obedience, and love expressed in everyday choices. *References: Romans 12:17–21; Philippians 2:14–16; Matthew 5:14–16.*

99. What is the Great Commission?

The Great Commission is Christ's command to His followers to make disciples of all nations, teaching them to obey everything He has commanded. This mission is empowered by the Holy Spirit and grounded in Christ's authority. The Great Commission is not optional; it is the calling of the entire Church. Believers participate in God's redemptive work by proclaiming the gospel and forming communities of faith. *References: Matthew 28:18–20; Acts 1:8; Romans 10:14–17.*

100. What is the Christian hope?

The Christian hope is the confident expectation of resurrection, renewal, and eternal life with God. It is grounded in the resurrection of Jesus Christ and the promises of Scripture. Hope looks forward to a restored creation where righteousness dwells, where sin and death are no more, and where God is all in all. This hope sustains believers in suffering and shapes their life in the present. Hope is not wishful thinking but trust in the unchanging character of God. *References: 1 Peter 1:3–5; Revelation 21:1–5; 1 Corinthians 15:20–28.*

PART NINE
CONCLUSION

CHAPTER TWENTY-FOUR
THE UNIVERSE AS A PERSONAL INVITATION

From the beginning of this book, we have traced a universe that is intelligible, relational, ordered, and filled with meaning. We have explored physics that reveals a world structured by information and interaction, metaphysics that shows why existence is not accidental, and theology that points to the God who calls Himself "I AM." We have journeyed from the edge of the cosmos to the foundations of Western civilization, from the nature of consciousness to the moral structure of reality, from the Trinity to the Sermon on the Mount. All of these threads lead to a single truth: the universe is not merely something to observe. It is an invitation.

The universe is not a cold machine or a lonely expanse drifting toward nothingness. It is a place where order reflects intention, where rationality reflects the Logos, and where consciousness awakens to relationship. The intelligibility of the world itself suggests that the One who made it is personal. A universe grounded in rational order is a universe grounded in purpose, and purpose belongs to someone. The order of the world is the echo of the One who speaks it into being. Scripture begins with the words, "In the beginning God created" (*Genesis 1:1, ESV*). These words are not only historical but relational. Creation is God's first act of generosity. The physical universe is not a barrier between God and humanity but the environment in which He makes Himself known.[1]

Modern physics reinforces this. The laws of nature are elegant. The universe is finely tuned for life. The arrow of entropy moves from simplicity toward complexity and awareness. In the (IAM), this does not mean the universe learns or that information is conscious. It means that God created a world in which rational beings can discover His wisdom. To exist in such a universe is already to be addressed by God. The very structure of reality invites investigation, wonder, and worship. Every equation that describes nature, every symmetry that physicists uncover, every law that governs the behavior of matter and energy—all of these point toward the rationality of the One who established them. The universe is not silent. It speaks. And what it speaks of is the character of its Creator.[2]

You are not an accident. You are not a cosmic fluke, the random byproduct of impersonal forces grinding through an indifferent universe. You are a person, made in the image of God. Human beings are the only creatures who ask why, who imagine what is not yet, and who reflect on their own thoughts. Scripture teaches that humanity is made in the image of God (*Genesis 1:26–27, ESV*). Physics cannot describe the soul, but it reveals that consciousness is not reducible to matter. Minds integrate information, reflect on themselves, and desire truth. Consciousness arises in a universe that is structured to support it, that is open to novelty, and that exhibits the kind of complexity necessary for awareness to emerge. This is not an accident. It is a reflection of the fact that the universe is grounded in the God who is personal, rational, and relational.[3]

In a universe grounded in the divine "I AM," personhood is part of the design. You can know God because you were made to know Him. You can love because you were created by the God who is love. Your capacity for reason reflects the Logos through whom all things were made. Your longing for meaning is not a psychological weakness or a social construct. It is a sign of the soul's origin. Only the I AM can reveal who you truly are. Every person is known fully and loved completely by God. Identity is not self-invention. It is the discovery of who God created you to be. And the invitation of God is written into the human soul.

The clearest revelation of God is not mathematical beauty or cosmic order but a person. Scripture teaches that the Word through whom all things were made became flesh (*John 1:14, ESV*). In Jesus Christ, the Creator entered His creation. The One who sustains every quantum state lived in human history and bore human sorrow. The Logos who spoke the universe into existence became a man, walked among us, taught, healed, suffered, died, and rose again. The incarnation reveals God's relational purpose. The universe shows that God is powerful. Christ shows that God is near. The resurrection proves that the One who founded the universe also redeems it. Christ is the final proof that the universe is personal at its core. Christ is the heart of the invitation.[4]

Freedom is one of the clearest signs of divine intention. Freedom allows for love, creativity, moral choice, and relationship. It also allows for sin and suffering. Scripture presents humanity as creatures capable of choosing God or rejecting Him. Freedom is necessary for love. A relationship without freedom is not a relationship. The universe bears the imprint of a God who desires freely chosen love. In the (IAM), freedom is not randomness. It is real participation in the purposes of God. The invitation of God respects human freedom because love cannot be forced. You are not a puppet. You are not programmed. You are a person, capable of genuine choice, and your choices matter. They matter because they shape your character, they affect others, and they determine the direction of your life. They matter because God has chosen to make your response part of the story He is telling through history.[5]

Every person knows that the world is broken. Pain, guilt, and longing reveal that something is not as it should be. Scripture explains this fracture as the result of sin. Sins that include the misuse of freedom, the rejection of God's relational order, the choice to turn away from the good. And Scripture reveals God's plan to heal it. Christ's death and resurrection are the center of that plan. Redemption is the restoration of what was lost and the reconciliation of humanity with God. The invitation of redemption is open to every person. No sin is too great, no past too dark. The God who calls Himself "I AM" also says, "Come to Me" (*Matthew 11:28, ESV*). He does not merely repair creation. He renews it. He does not merely forgive sin. He transforms the sinner. He does not merely offer a second chance. He offers a new life.[6]

The Christian hope is not escape from the world but the renewal of all things. Scripture promises a new heaven and a new earth (*Revelation 21:1, ESV*). In that world, God's presence is fully revealed. Death and sorrow are gone. Eternity is the fulfillment of the invitation that began with creation and continued through redemption. The universe points toward this final reality. Creation is relational at its core, sustained by the God who calls all things into being. Eternity is the completion of that calling. It is not an otherworldly escape but the restoration of this world, the healing of all that is broken, and the full realization of the communion for which we were made.

The greatest truth this book offers is not merely that the universe is structured and ordered. It is that the God who sustains the universe knows you, loves you, and invites you into relationship. The call of I AM is personal. He speaks to you by name. Your existence is not an accident. Your consciousness is not an illusion. Your longing for meaning is real. You were made to know the One whose voice spoke reality into being. The universe is God's invitation. Christ is God's answer. Your response is the final movement in the story of existence.[7]

Salvation is a gift that is made available to those who repent, believe, and confess that Jesus is Lord and that He died and rose from the dead to save humanity (*Acts 16:31; Romans 10: 9–10, ESV*). This gift cannot be earned through good deeds (*Ephesians 2:8–9; Titus 3:5, ESV*). It requires faith, acting on what you believe according to God's Word concerning salvation. Faith is not mere intellectual assent. It is trust, commitment, and surrender. It is the recognition that you cannot save yourself, that you need the grace of God, and that you are willing to follow Christ wherever He leads. It is the decision to build your life on the foundation of the One who says "I AM," to receive the identity He offers, to pursue the purpose He reveals, and to trust that the One who made you will also sustain you, guide you, and bring you home.[8]

If anything in this book has stirred more than curiosity. If you began to feel the questions becoming personal, then the next move is to respond to that calling. For some readers, that response will be quiet: a first prayer, a return to prayer, or simply a willingness to ask God to be known. Prayer does not need to be eloquent. It needs to be honest. You can speak to God as you would speak to someone you trust, someone who knows you completely and loves you still. You can confess your doubts, your fears, your failures, and your longing. You can ask Him to reveal Himself to you, to make His presence known, to give you the faith to follow. And He will answer. He promises that those who seek will find, that those who knock will have the door opened, that those who ask will receive (*Matthew 7:7–8, ESV*).[9]

For others, the response will mean seeking a church community and beginning to learn the faith from the inside, not as an idea, but as a lived reality. Christianity is not a philosophy to be studied at a distance. It is a way of life to be practiced in community. The church is not a building or an institution. It is the body of Christ, the gathering of people who have responded to the invitation of the I AM and who are learning together what it means to follow Him. In the church, you will find imperfect people trying to live in alignment with God's character. You will find worship, teaching, fellowship, and service. You will find brothers and sisters who can encourage you, challenge you, pray for you, and walk with you. You will find sacraments, baptism and communion, that mark your identity as one who belongs to Christ and that remind you of His grace. Finding a church is not optional. It is essential. We were not made to follow Christ in isolation. We were made for relationship, and the church is where we learn to live in the relational order God established.[10]

Practical next steps include reading Scripture regularly. The Bible is not merely a historical document or a collection of religious teachings. It is the Word of God, the means by which He speaks to His people. Start with the Gospels Matthew, Mark, Luke, or John. Read slowly. Pay attention to the words of Jesus. Notice how He treats people, how He speaks of God, how He calls His followers to live. Let the text shape your imagination. Let it challenge your assumptions. Let it reveal the character of the One who made you. As you read, pray. Ask God to help you understand. Ask Him to show you how His Word applies to your life. And trust that He will.

Pray consistently. Prayer is not a ritual or a formula. It is conversation with God. It is bringing your whole self; your joys, your sorrows, your questions, and your gratitude before the One who knows you completely. Prayer changes you. It does not change God, but it changes your heart. It aligns your will with His. It opens you to His presence. It reminds you that you are not alone. Set aside time each day, even if it is only a few minutes. Speak to God honestly. Listen for His voice in Scripture, in the counsel of wise believers, in the quiet conviction of the Holy Spirit. And trust that He hears you.

Learn from mature believers. Christianity has been practiced for two thousand years. Countless men and women have walked the path of faith before you. They have wrestled with the same questions, faced the same doubts, experienced the same joys. Their wisdom is available to you. Find someone who has walked with Christ for many years and ask them to mentor you. Read the works of faithful teachers like Augustine, Athanasius, Luther, Calvin, Edwards, Lewis, Stott, Willard. These are not infallible guides, but they are brothers and sisters who have devoted their lives to understanding and living the faith. Learn from them. Let their insights deepen your own understanding.

Serve others. Christianity is not a private spirituality. It is a call to love God and love your neighbor. Look for ways to serve. Whether in your church, in your neighborhood, or in your workplace. Serve those who cannot repay you. Care for the poor, the sick, the lonely, the marginalized. When you serve, you reflect the character of Christ, who came not to be served but to serve (*Mark 10:45, ESV*). And when you serve, you discover that the life God calls you to is not burdensome but life-giving. You find that in giving, you receive. In serving, you are served. In loving, you are loved.

Be patient with yourself. The Christian life is not a single decision followed by moral perfection. It is a journey. It is the slow transformation of the heart, the gradual conforming of your character to the image of Christ. You will stumble. You will fail. You will doubt. But God does not abandon you. He sustains you. He forgives you. He picks you up and sets you back on the path. Do not measure your progress by the standards of this world: success, achievement, status. Measure it by faithfulness: Are you growing in love? Are you pursuing truth? Are you learning to trust God even when you do not understand? That is enough. That is more than enough.

The universe is an invitation. Christ is the answer. And the door is open. "Behold, I stand at the door and knock. If anyone hears my voice and opens the door, I will come in to him and eat with him, and he with me" (*Revelation 3:20, ESV*). The God who sustains every star, who holds every atom in being, who knows every thought and counts every tear invites you to communion with Him. He does not force. He does not coerce. He knocks. And He waits. The universe is His invitation. Christ is His voice. And your response, whatever it is, will shape not only your life but the lives of those around you, the communities you belong to, and the world you help create.

Open the door. Respond to the invitation. And discover that the One who made you has been calling you home all along.[11]

CHAPTER REVIEW:
KEY CONCEPTS:

- **The universe reveals a personal God** Not just abstract principles but the One who says "I AM"

- **Christ as God's answer** Logos made flesh; divine response to human condition

- **Salvation through faith** Grace received, not earned; repentance and belief

- **Response required** Knowledge demands response; invitation awaits answer

- **Practical steps** Prayer, Scripture, community, baptism, ongoing discipleship

WHAT THIS MEANS:

Everything the book has developed—physics, philosophy, theology—points toward this: the God who created and sustains all reality invites you into relationship. This is not abstract theology but personal invitation. The appropriate response is not just intellectual assent but life reorientation through faith in Christ.

IMPORTANT TERMS & CONCEPTS:

- **Invitation (theological)** → Throughout conclusion
- **Salvation** → Part X, Q&A Section VI
- **Faith** → Q&A Section VI; Appendix A
- **Repentance** → Appendix A (prayers); Q&A Section VI
- **Discipleship** → Q&A Section X

LOOKING AHEAD:

- Appendix A provides specific prayers from different traditions
- Q&A Section X gives practical guidance
- Bibliography and Glossary provide further study resources
- Part X remains available for ongoing reference

TECHNICAL REFERENCE:

- **For salvation theology:** Q&A Section VI (Questions 51-60)
- **For prayer resources:** Appendix A
- **For next steps:** Appendix A and Q&A Section X (Questions 91-100)
- **For denominational differences:** Conclusion endnotes; Appendix A notes

---

The Universe as a Personal Invitation 325

1. For the opening of Genesis as both historical and relational, see: *Genesis 1:1; John 1:1–3*. Barth, K. (1958). *Church Dogmatics III/1: The Doctrine of Creation*. T&T Clark, §41. Creation is God's first act of grace, the free gift of existence to what is not God, establishing relationship from the beginning.
2. For the fine-tuning of the universe and the intelligibility of natural law as pointing toward a rational Creator, see: Collins, R. (2009). "The Teleological Argument: An Exploration of the Fine-Tuning of the Universe." In *The Blackwell Companion to Natural Theology*, edited by W. L. Craig and J. P. Moreland. Wiley-Blackwell, pp. 202–281. Wigner, E. (1960). "The Unreasonable Effectiveness of Mathematics in the Natural Sciences." *Communications in Pure and Applied Mathematics* 13(1): 1–14. The mathematical structure of nature suggests a rational source.
3. [3] For the biblical doctrine of the image of God and its implications for human consciousness and personhood, see: *Genesis 1:26–27; Psalm 8:3–8*. Theological treatment: Hoekema, A. A. (1986). *Created in God's Image*. Eerdmans, Chapters 1–4. Middleton, J. R. (2005). *The Liberating Image: The Imago Dei in Genesis 1*. Brazos Press. Bearing God's image entails rational capacity, moral agency, creativity, and relationality.
4. For the incarnation as the ultimate revelation of God's character and the climax of His relational purpose, see: *John 1:14, 18; Colossians 1:15–20; Hebrews 1:1–3*. Torrance, T. F. (1992). *The Mediation of Christ*, revised edition. Helmers & Howard, Chapters 2–3. Barth, K. (1956). *Church Dogmatics IV/1: The Doctrine of Reconciliation*. T&T Clark, §59. The incarnation reveals that God is not remote but near, not abstract but personal.
5. For freedom as necessary for love and relationship, see: Swinburne, R. (1998). *Providence and the Problem of Evil*. Oxford University Press, Chapters 5–6. Plantinga, A. (1974). *God, Freedom, and Evil*. Eerdmans. God grants genuine freedom because love, to be real, must be freely chosen; this is why the universe exhibits openness and indeterminacy.
6. For redemption as restoration and renewal, see: *2 Corinthians 5:17–21; Romans 8:18–25; Revelation 21:1–5*. Wright, N. T. (2008). *Surprised by Hope: Rethinking Heaven, the Resurrection, and the Mission of the Church*. HarperOne, Chapters 1–6, 11. Christian hope is not escape from creation but its transformation and renewal.
7. For God's personal call and the relational nature of salvation, see: *John 10:3, 14; Isaiah 43:1*. Barth, K. (1956). *Church Dogmatics IV/1: *The Doctrine of Reconciliation**. T&T Clark, §58 (The Doctrine of Reconciliation). God's call is specific, personal, and addressed to each individual by name.
8. For the biblical doctrine of salvation by grace through faith, see: *Ephesians 2:8–9; Romans 3:21–26; Acts 16:30–31; Romans 10:9–13; Titus 3:4–7*. Stott, J. R. W. (1986). *The Cross of Christ*. InterVarsity Press, Chapters 7–9. Luther, M. (1520). *The Freedom of a Christian*. Salvation is received by faith, not earned by works, though saving faith produces works as its fruit.
9. [9] For prayer as honest conversation with God and the promise that those who seek will find, see: *Matthew 7:7–11; Philippians 4:6–7; 1 Thessalonians 5:17*. Foster, R. J. (1992). *Prayer: Finding the Heart's True Home*. HarperSanFrancisco, Chapters 1–3. Prayer is not ritual but relationship, bringing one's whole self before God.

10. For the necessity of the church as the community of believers, see: *Hebrews 10:24–25; 1 Corinthians 12:12–27; Ephesians 4:11–16.* Bonhoeffer, D. (1954). *Life Together.* Harper & Row. Clapp, R. (1996). *A Peculiar People: The Church as Culture in a Post-Christian Society.* InterVarsity Press. The Christian life is inherently communal; faith is lived in fellowship, not in isolation.
11. For *Revelation 3:20* and the open invitation of Christ, see: *Revelation 3:20; John 6:37.* Lewis, C. S. (1952). *Mere Christianity.* Macmillan, Book IV, Chapter 8. The call of Christ is an invitation, not coercion; He stands at the door and knocks, waiting for the door to be opened from within.

PART TEN

THE TECHNICAL REFERENCE COMPANION & INTERDISCIPLINARY CLARIFICATION MANUAL

SECTION SUMMARIES

SECTION 1 Summary: Domain Boundaries & Methodological Clarifications

Modern cosmology, classical philosophy, and Christian theology each speak with their own vocabulary and methods. When the same word—like 'information,' 'order,' or 'cause'—appears in multiple disciplines, it often means different things. This section clarifies what each discipline actually studies and where its boundaries lie.

Physics describes mechanisms and observable behavior. Metaphysics examines being, causality, and the conditions that make existence possible. Theology addresses ultimate origin, divine action, and meaning. Information theory quantifies distinctions and correlations. Philosophy of mind explores consciousness and subjectivity.

Understanding these boundaries prevents confusion and allows genuine dialogue. The unified vision this work presents is interpretive coherence—showing how these perspectives illuminate the same reality from different angles—not ontological collapse where one discipline becomes another.

When you encounter terms that appear in multiple chapters, refer back to this section to see which domain is being discussed. This prevents category errors and ensures you're engaging with what the book actually claims, not assumptions imported from elsewhere.

A Note on the Aristotelian-Thomistic Framework

You may notice that this work relies heavily on Aristotelian-Thomistic metaphysics—the philosophical framework developed by Thomas Aquinas in synthesizing Aristotle with Christian theology. This isn't arbitrary traditionalism.

Aquinas faced a challenge similar to ours: integrating the best available natural philosophy (Aristotle's physics) with Christian faith. He recognized that metaphysics—specifically, Aristotle's categories of being, causality, potency, and act—provided the necessary conceptual tools to bridge the two without collapsing them into each other.

That synthesis didn't just produce academic philosophy. It gave intellectual structure to Western concepts of natural law, human dignity, and inalienable rights that shape free societies today.

This book follows the same methodological path because the structure of reality demands it. Modern cosmology (like Aristotle's physics) describes how the universe behaves. Christian theology (then and now) describes who God is. Metaphysics provides the bridge by asking questions neither science nor theology alone can answer: Why does anything exist? What is being? What grounds contingent reality?

Aristotelian-Thomistic philosophy isn't the only metaphysical system, but it remains the most rigorously developed framework for this kind of synthesis—which is why it's still taught in university philosophy departments worldwide, 800 years after Aquinas.

You're not being asked to become a medieval philosopher. You're being invited to use the same intellectual tools that have proven, over centuries, to be the most honest and rigorous way to hold science and faith together.

SECTION 2 Summary: Category Errors & Domain Separation Principles

Category errors happen when concepts from one discipline get misapplied in another—like treating physical laws as metaphysical necessities, interpreting quantum 'observers' as conscious minds, or equating information with meaning.

This section provides clear operational definitions for each domain's proper scope. The principle of non-reduction is simple: physics cannot reduce metaphysics, consciousness cannot be exhaustively explained by computation, and divine action operates on a different explanatory level than natural causation.

Why this matters for you: When reading this book, you'll encounter terms like 'information,' 'potential,' and 'order' in different contexts. This section helps you track which domain is speaking at any given moment, preventing the confusion that derails most interdisciplinary conversations.

The distinction between primary causation (metaphysical/theological: why anything exists) and secondary causation (physical: how things interact) follows classical theological treatment that has guided Christian thinkers for centuries. Recognizing these boundaries isn't intellectual rigidity—it's methodological honesty.

SECTION 3 Summary: What This Work Does Not Claim

Before explaining what this work does claim, it's essential to be clear about what it does not claim. Interdisciplinary synthesis invites predictable misreadings, so this section explicitly addresses them.

This work does not endorse pantheism, panentheism, or panpsychism. It does not claim consciousness causes quantum outcomes or that observation by a mind determines physical reality. It does not equate information with consciousness, divinity, or meaning. It does not assert the universe is a simulation or a literal hologram. It does not reduce theology to physics or physics to theology. It does not treat God as identical with natural processes, energy, or information. And it does not claim physics proves metaphysics or theology.

What this work offers instead: Structural resonances and conceptual coherence. When modern cosmology reveals an ordered, intelligible, relational universe, and Christian theology describes a God who is Logos (divine Reason), the resonance is worth noting—not as proof, but as coherence.

This section prevents straw-man critiques and ensures you engage with the actual argument being made, not a distorted version of it.

SECTION 4: Common Misunderstandings & Clarifications

As you read this book, certain questions will naturally arise: 'Does quantum mechanics mean consciousness affects reality?' 'Is God just another word for the laws of nature?' 'Doesn't information theory prove the universe is a computer?' These are good questions, and this section addresses them directly.

Part A: Questions Most Readers Have addresses the fifteen most common confusions that arise when science, philosophy, and theology intersect. If you're reading this book for the first time, focus on Part A—it clarifies exactly what the synthesis does and doesn't mean.

Part B: Advanced Technical Distinctions addresses specialized philosophical terminology for readers familiar with academic debates in metaphysics or philosophy of mind. Most readers can skip Part B on first reading.

The goal isn't to overwhelm you with technicalities—it's to ensure you understand what's actually being claimed so you can evaluate it fairly. When you encounter a term or concept that confuses you in the main chapters, this is where you check for clarification.

SECTION 5 Summary: Anticipated Objections & Scholarly Responses

This section is for advanced readers and can be skipped entirely without loss of understanding.

If you're engaging with this work in an academic context, teaching it in a university setting, or simply interested in how the framework responds to specialized critiques from physics, philosophy of mind, metaphysics, or theology, this section provides structured responses.

Most readers will find everything they need in Sections 1-4 and 6-7. This section exists for completeness and for readers who want to examine how the synthesis holds up under rigorous scrutiny from multiple disciplines.

The objections are stated sharply (as critics would actually phrase them) followed by precise academic clarifications. This is not essential reading for understanding the book's argument—it's optional material for those who want to go deeper into methodological questions.

SECTION 6 Summary: Cross-Discipline Translation Tables

This might be the most practically useful section in Part X.

When the same word appears in physics, philosophy, and theology, it almost always means something different. 'Information' in physics (measurable states) is not 'information' in biology (genetic encoding) is not 'information' in theology (divine knowledge). 'Observer' in quantum mechanics (any measuring system) is not 'observer' in philosophy (conscious subject).

These translation tables map how key terms function across disciplines, preventing the confusion that collapses most interdisciplinary conversations. Think of this as your conceptual dictionary—when you encounter a term used in different ways across chapters, come here to see the precise distinctions.

How to use these tables: When reading and you notice a word appearing in multiple contexts (like 'cause,' 'order,' 'being,' 'potential'), check the relevant table to see which domain's meaning applies in that passage. This single tool prevents most category errors and ensures you're tracking the argument accurately.

SECTION 7 Summary: Integration Guidelines

How do you think across disciplines without collapsing them into each other? This section provides eleven explicit principles for combining insights from different fields while preserving their boundaries.

These aren't just rules for THIS book—they're principles you can apply to any interdisciplinary work:

Complementarity – Different disciplines offer complementary explanations, not competing ones

Non-Substitution – No discipline can replace another

Hierarchical Explanation – Reality has layers; higher levels don't override lower mechanisms

Non-Reduction – Complex phenomena can't be reduced to single explanatory frameworks

...and seven more that show how physics, metaphysics, philosophy of mind, information theory, and theology can speak together without any one dominating or replacing the others.

Think of these as intellectual guardrails. They keep the synthesis honest, prevent overreach, and ensure each discipline contributes what it does best. When you encounter a passage where multiple disciplines interact, these principles show you how to evaluate whether the integration is legitimate or whether a boundary has been crossed.

SECTION 8 Summary: Comparative Frameworks

If you're familiar with other attempts to relate science and religion—like process theology, digital physics, panpsychism, or quantum mysticism—you might wonder: 'How is this different?'

This section explicitly compares the (IAM) framework to other modern theories that might appear similar at first glance, showing precisely where they align and where they diverge.

Understanding these distinctions prevents misclassification and ensures this work is engaged on its own terms: it is a synthesis of classical theism, Aristotelian-Thomistic metaphysics, contemporary physics, and information theory that respects disciplinary boundaries while identifying structural resonances.

It is neither digital physics (which treats the universe as literally computational), nor panpsychism (which attributes consciousness to matter), nor process theology (which revises classical divine attributes), nor quantum mysticism (which misapplies quantum mechanics to consciousness). It is a carefully bounded interdisciplinary framework grounded in methodological rigor and philosophical precision.

For most readers: Skim the table of comparisons to see the landscape. For scholars: This section demonstrates exactly where (IAM) fits in contemporary science-religion dialogue.

SECTION 1
DOMAIN BOUNDARIES & METHODOLOGICAL CLARIFICATIONS

The purpose of this section is to clarify the boundaries between the major disciplines referenced throughout this book. Physics, metaphysics, theology, and information theory often employ overlapping language—terms such as "order," "cause," "information," "observer," and "being" appear in multiple contexts but with different meanings. Without explicit boundary-setting, readers may mistakenly assume that these disciplines speak in identical conceptual categories. They do not.

This section establishes the methodological distinctions necessary for serious interdisciplinary work. It ensures that the arguments presented in this book are not misconstrued as conflating domains or making claims outside their rightful scope.

1.1 Physics and Its Domain of Explanation

Physics describes the structure, behavior, and laws governing the physical universe. Its primary aims are:

- Quantitative description of observable phenomena.
- Predictive models expressed in mathematical form.
- Identification of symmetries, constraints, and interactions.

- Empirical testability as a methodological criterion.

Physics does not address:

- metaphysical necessity
- ultimate causes
- why something exists rather than nothing
- meaning, value, or intentionality
- the nature of personhood
- the existence or attributes of God

Even when physics discusses "laws," these are descriptive regularities, not metaphysical necessities. When physics uses terms like "information," "observer," or "potential," these terms refer to formal structures or interactions, not philosophical or psychological realities.

1.2 Metaphysics and Its Domain of Explanation

Metaphysics investigates the foundational principles of reality as such. Its primary aims are:

- To identify what must be true for anything to exist.
- To distinguish between kinds of being (necessary vs. contingent).
- To explain causality, identity, essence, and possibility.
- To examine the structure of intelligibility and the act of being.

Metaphysics does not:

- function as physics under another name
- make empirical predictions
- describe physical mechanisms
- replace scientific explanation

Metaphysical explanations are ontological: they reveal the

conditions under which scientific explanations themselves become possible.

1.3 Theology and Its Domain of Explanation

Theology studies God, divine action, revelation, and humanity's relationship to the divine. Its primary aims include:

- To articulate the nature and attributes of God.
- To examine the relationship between Creator and creation.
- To interpret revelation (Scripture, tradition, reason).
- To explore moral, existential, and spiritual truths.

Theology does not:

- replace physics
- treat God as an element of the physical universe
- propose natural mechanisms for divine action
- reduce revelation to metaphor or physics to symbolism

When theology speaks of "order," "cause," "law," "word," or "spirit," these terms refer to personal and metaphysical realities, not physical processes

1.4 Information Theory and Its Domain of Explanation

Information theory studies:

- The representation of distinctions.
- The transmission and transformation of signals.
- Quantitative uncertainty (Shannon information).
- Structural complexity and constraints.
- The informational architecture underlying systems.

Information theory does not:

- define meaning
- describe consciousness

- explain existence
- address metaphysical necessity
- simply intelligence or intentionality

When this book connects information and metaphysics, it does so analogically and structurally, not by equating information with intelligence or awareness. Information theory provides a descriptive framework for discussing order, distinctions, and structure—not a causal mechanism. The universe can be described using information-theoretic language without claiming that information causes or drives physical processes.

1.5 How These Domains Interrelate

The four domains interact without collapsing into one another:

- Physics provides the formal, measurable structure of the universe.
- Information theory describes the formal architecture through which physical distinctions are encoded.
- Metaphysics explains being, act, and the conditions for intelligibility.
- Theology describes the ultimate ground of being, intelligibility, and relationality.

Each domain has its own proper integrity. When they speak to each other, they do so through:

- analogy
- correspondence
- structural resonance
- philosophical interpretation

NOT through forced equivalence.

1.6 Why Boundary Clarifications Are Necessary

A single misunderstood term can create the appearance of:

- pseudoscience
- category errors
- theological confusion
- philosophical incoherence
- scientific overreach

This book avoids these dangers by:

- Using each term within its disciplinary meaning.
- Marking analogies explicitly.
- Separating empirical claims from metaphysical ones.
- Clarifying theological uses from scientific uses.
- Providing this manual to make disciplinary distinctions explicit.

1.7 The Unified Vision Without Category Collapse

The (IAM) argues that:

- physics reveals a structured, relational, informational universe
- metaphysics reveals a grounded act of being
- theology reveals a personal ground of intelligibility
- consciousness participates in this relational structure

But it does not:

- turn physics into theology
- turn theology into physics
- turn metaphysics into cosmology
- turn information into intelligence
- turn quantum potentiality into consciousness

The unity is conceptual and interpretive, not ontological collapse.

SECTION 2
CATEGORY ERRORS AND DOMAIN SEPARATION PRINCIPLES

Category errors occur when concepts from one domain (physics, metaphysics, theology, information theory, or philosophy of mind) are misapplied in another. This section establishes strict boundaries to prevent such errors and outlines how each discipline should be understood within its own methodological framework.

2.1 Definition of Category Error

A category error occurs when:

- a concept belonging to one domain is treated as if it belongs to another,
- a term with multiple meanings is used without specifying its domain, or
- an explanation proper to one discipline is mistaken for an explanation in another.

Classic examples include:

- treating physical laws as metaphysical necessities
- interpreting quantum "observers" as conscious minds

- equating divine causality with physical causation
- interpreting metaphysical "act" as physical activity
- treating "information" as equivalent to "meaning" or "consciousness"

This section prevents such misunderstandings.

2.2 The Five Major Domains & Their Proper Questions

Each domain answers fundamentally different questions.

Physics

Questions addressed:

- How do physical systems behave?
- What are the laws and structures governing matter, energy, and spacetime?
- What predictions can be tested empirically?

Questions not addressed:

- Why does anything exist at all?
- What is the nature of consciousness?
- What is the ground of being?

Metaphysics

Questions addressed:

- What does it mean to exist?
- What are the principles underlying being, causality, and identity?
- What distinguishes contingent from necessary existence?

Questions not addressed:

- What is the mass of the electron?
- How do physical forces operate?

- What mechanisms cause observable phenomena?

Theology
Questions addressed:

- Who is God?
- What is the relationship between Creator and creation?
- What is meaning, purpose, and final destiny?

Questions not addressed:

- What are the equations governing cosmology?
- How do quantum fields interact?
- What are the empirical mechanisms underlying nature?

Information Theory
Questions addressed:

- How are distinctions encoded and transmitted?
- How much uncertainty exists in a system?
- What is the structure of messages and signals?

Questions not addressed:

- What is consciousness?
- What is metaphysical or divine knowledge?
- What is the nature of existence

Philosophy of Mind
Questions addressed:

- What is consciousness?
- What is the nature of mental states?
- How do subjectivity and intentionality function?

Questions not addressed:

- What is the structure of a quantum field?
- What is the metaphysical act of being?
- What is divine providence?

2.3 Why Category Errors Happen in Interdisciplinary Work
Category errors commonly arise when:

- the same word is used in different disciplines (e.g., "information," "cause," "observer," "order," "purpose," "potential"),
- analogical language (especially theological language) is misunderstood as literal,
- metaphysical concepts are interpreted as physical mechanisms,
- scientific descriptions are mistakenly seen as metaphysical explanations,
- physical models are treated as ultimate explanations of reality.

This section ensures such mistakes are systematically avoided.

2.4 The Principle of Non-Reduction

- No domain should be reduced to another.
- Physics cannot reduce metaphysics.
- Physics describes how the universe behaves, not why it exists.
- Metaphysics cannot replace physics.
- Metaphysics describes being and causation at a deeper level but does not describe mechanisms.
- Theology does not replace physics or metaphysics.
- Theology concerns God and divine action, not physical processes.

- Information theory does not describe meaning or consciousness.
- It quantifies distinctions; it does not generate or explain interior experience.
- Consciousness cannot be reduced to computation or physical states.
- Subjective awareness cannot be equated with algorithmic operations or brain activity.

These boundaries preserve intellectual integrity.

2.5 The Principle of Analogical Language

Many terms in theology and metaphysics are analogical, not literal or physical.

Examples:

- "God is light" does not mean electromagnetic radiation
- "Word" (Logos) does not mean linguistic grammar
- "Wisdom" (Chokhmah) is not computational intelligence
- "Act" (actus essendi) is not kinetic motion
- "Cause" (primary cause) is not a physical signal

Analogical language must be interpreted within its proper domain. Failing to recognize analogy is a major source of category confusion.

2.6 The Principle of Domain-Specific Causality

Each discipline uses the word cause differently:

Physical cause:
Interaction described by equations (forces, fields, dynamics).

Metaphysical cause:
The explanation of a thing's existence or nature.

Theological cause:
God as the sustaining source of being (primary cause).

Informational cause:
A constraint or difference that makes a difference.

Mental cause:
Intentional, rational, or volitional action.
These are not interchangeable.

2.7 The Principle of Layered Explanation

Reality requires multiple complementary explanations, each proper to its level.

Example: A written letter can be explained by:

- Physics: ink patterns on paper
- Information theory: encoding of symbols
- Linguistics: structure of language
- Philosophy of mind: intention of the author
- Theology/metaphysics: existence and rationality of the author

No single level replaces the others.

2.8 Domain Separation Table (Quick Reference)

Term	Physics	Metaphysics	Theology	Info Theory	Mind
Cause	Force, dynamics	Act of being	Divine action	Informational constraint	Intention
Order	Symmetry	Teleology	Providence	Structure	Ratianality
Observer	Measuring system	Subject of being	Personal agent	Struciver	Conscious subject
Information	Physical state	Intelligibility	Divine knowledge	Shannon data	Meaning
Potential	Probability amplitude	Potency	Possible action	Data capacity	Intentional stare
Law	Mathematical regularity	Principle of buing	Divine command/order	Encoding rule	Norm

Domain Separation Table (Quick Reference)

SECTION 3
WHAT THIS WORK DOES NOT CLAIM

This section establishes explicit denials of positions that are commonly misattributed to interdisciplinary writings on physics, metaphysics, and theology. The following clarifications guard against category errors, overextensions, and misinterpretations of terminology. Each point is stated directly, without nuance, so that no reader can mistakenly impose views that this work does not endorse or imply.

3.1 No Equating of God with Natural Processes
This work does not claim that:

- God is identical with the universe
- God is identical with physical laws
- God is identical with quantum fields
- God is identical with energy
- God is identical with information
- God is identical with emergent complexity

God, in classical theism, is transcendent, personal, simple, and the ground of all being, not a component or phase of the natural world.

3.2 No Pantheism, Panentheism, or Panpsychism

This work does not endorse:

- pantheism (the universe is God)
- panentheism (the universe is within God as part of God's being)
- panpsychism (all matter possesses consciousness)
- cosmopsychism (the universe as a whole is conscious)

None of the analogies drawn between physical structure and metaphysical intelligibility imply that consciousness or divinity is intrinsic to matter.

3.3 No Claim That Consciousness Causes Quantum Outcomes

- This work does not claim that:
- consciousness collapses the wave function
- observation by a mind determines physical reality
- human awareness influences quantum states
- subjective experience has physical control over particles

Quantum measurement involves physical interactions, not mental ones. Nothing in this work relies on or assumes consciousness-induced wave function collapse.

3.4 No Claim That Information Is Conscious, Divine, or Personal

This work does not claim that:

- information is a form of consciousness
- information is metaphysically ultimate
- information possesses intentionality
- information carries meaning in itself
- information is an attribute of God in a literal sense

Information, as used here, is a formal and structural concept, not a personal or spiritual one.

3.5 No Conflation of Information with Meaning or Intelligence

This work does not equate:

- Shannon information with semantic meaning
- structural distinctions with rationality
- physical signals with intentional thought
- data with wisdom

Meaning, intelligence, and rationality belong to persons, not to physical systems.

3.6 No Claim That Physics Proves Metaphysics or Theology

This work does not assert that:

- physics proves the existence of God
- quantum mechanics implies theism
- cosmic fine-tuning is a formal proof of divine action
- the laws of physics demonstrate metaphysical premises
- any scientific model validates theological doctrine

Physics can suggest or resonate with metaphysical ideas, but it does not demonstrate them.

3.7 No Sacredization of Physical Models

This work does not treat:

- the Big Bang
- inflation
- holography
- general relativity
- quantum field theory

as theological or metaphysical truths. Physical theories are models, not ultimate explanations.

3.8 No Claim That Metaphysics Predicts Physical Phenomena

This work does not claim that metaphysics:

- predicts empirical observations
- replaces scientific inference
- explains mechanisms or dynamics
- anticipates future discoveries in physics

Metaphysics explains being, not physical behavior.

3.9 No Reduction of Theology to Physics or Metaphysics
This work does not reduce:

- God to metaphysical abstraction
- divine action to physical causation
- providence to determinism
- revelation to natural processes
- the Logos to mathematical structure

Theology maintains its personal, relational, and revelatory nature.

3.10 No Oversimplification of Consciousness
This work does not claim that consciousness:

- emerges purely from complexity
- is identical to computation
- is reducible to neural states
- is caused by quantum events
- is an illusion created by brain processes

Consciousness is treated as a real, irreducible dimension of experience.

3.11 No Universal Mind, No Cosmic Awareness
This work does not suggest:

- the universe is conscious
- the universe is self-aware
- nature "learns" in a literal sense
- physical fields possess awareness

Analogical language remains analogical and is never intended as literal ontology.

3.12 No Misuse of Theological Language

This work does not interpret terms like:

- light
- word
- breath
- wisdom
- spirit

as metaphors for physical phenomena. Theological language is theological, not scientific.

3.13 No Claim That Human Knowledge Creates Reality

This work does not imply that:

- reality depends on human perception
- the world is a mental construct
- consciousness brings the universe into existence

Reality is objective and independent of human awareness.

3.14 No Claim That the Universe Is a Simulation or Projection

This work does not assert that:

- the universe is literally a hologram
- existence is a computer simulation
- consciousness projects reality
- reality is an illusion

Holography and information theory are mathematical frameworks, not metaphysical declarations.

3.15 No Collapsing of Domains

This work does not:

- turn physics into theology

- turn theology into physics
- turn information theory into metaphysics
- turn metaphysics into cosmology

Every field retains its own method, meaning, and authority.

SECTION 4
COMMON MISUNDERSTANDINGS AND CLARIFICATIONS

4.1 Misunderstandings Concerning Physics
(Questions Most Readers Have):

Misunderstanding 1:
"Quantum observers must be conscious minds."

Clarification:
In quantum mechanics, observer means any physical system that causes definite outcomes. Consciousness plays no necessary role.

Misunderstanding 2:
"Wave function collapse is caused by human observation."

Clarification:
Collapse is an interpretive term for an informational update triggered by physical interaction, not awareness.

Misunderstanding 3:
"Quantum indeterminacy implies metaphysical randomness."

Clarification:
Quantum probabilities are mathematical and structural; they do not describe metaphysical unpredictability or ontological chaos.

Misunderstanding 4:
"Holography means the universe is an illusion."

Clarification:

The holographic principle is a mathematical relationship between boundary and bulk descriptions, not a claim that reality is unreal.

(Advanced Technical Distinctions):

Misunderstanding 5:

"Entanglement means mystical connection or psychic unity."

Clarification:

Entanglement consists solely of mathematical correlations. It does not allow communication, intention, or consciousness between systems.

Misunderstanding 6:

"Fine-tuning proves design or disproves randomness."

Clarification:

Fine-tuning is an empirical observation; its interpretation requires metaphysics or theology, not physics.

Misunderstanding 7:

"Dark energy or vacuum energy is spiritual or metaphysical."

Clarification:

These are purely physical constructs with no inherent metaphysical or theological meaning.

Misunderstanding 8:

"Inflation explains the origin of the universe."

Clarification:

Inflation explains early cosmic conditions, not the metaphysical origin of existence or the reason for being.

4.2 Misunderstandings Concerning Metaphysics

(Questions Most Readers Have):

Misunderstanding 9:

"Metaphysics competes with scientific explanation."

Clarification:

Metaphysics explains existence, identity, and causality at a level that underlies physics; it does not replace empirical science.

Misunderstanding 10:

"Teleology implies conscious intention."

Clarification:
Metaphysical finality refers to directedness toward outcomes, not mental purpose or design.

Misunderstanding 11:
"Necessary being means a being that lasts forever in time."

Clarification:
Necessity is ontological, not temporal. A necessary being cannot not exist; time is irrelevant to its mode of being.

(Advanced Technical Distinctions):

Misunderstanding 12:
"Potentiality in metaphysics is the same as quantum potential."

Clarification:
Metaphysical potency concerns what a being can become. Quantum potential describes probability amplitudes. They are unrelated.

Misunderstanding 13:
"Metaphysical act is physical activity."

Clarification:
Metaphysical act refers to the actuality of existence, not physical motion or energetic change.

Misunderstanding 14:
"Being itself is a physical substrate."

Clarification:
Being itself is not a physical state, field, or substance. It is the most fundamental metaphysical principle.

4.3 Misunderstandings Concerning Information Theory

(Questions Most Readers Have):

Misunderstanding 15:
"Information is a metaphysical essence."

Clarification:
Information is formal, not ontological. It does not exist as a substance or metaphysical entity.

Misunderstanding 16:
"Landauer's principle shows information is physical being."

Clarification:

Landauer connects computation and thermodynamics, not metaphysics and ontology.

Misunderstanding 17:
"Semantic meaning is encoded in physical states."
Clarification:
Semantics requires a mind; physics handles only syntactic structures.

(Advanced Technical Distinctions):

Misunderstanding 18:
"Information in physics carries meaning or intelligence."
Clarification:
Physical information is structural and quantitative; it has no semantic or intentional content.

Misunderstanding 19:
"Entropy and information are the same thing."
Clarification:
In standard physics, entropy is a measure of disorder or uncertainty.

4.4 Misunderstandings Concerning Consciousness

(Questions Most Readers Have):

Misunderstanding 20:
"Consciousness emerges from complexity automatically."
Clarification:
No scientific explanation links physical complexity to subjective experience. Emergence describes structure, not awareness.

Misunderstanding 21:
"The brain generates consciousness as a physical product."
Clarification:
Neural activity correlates with consciousness but does not explain subjective awareness.

(Advanced Technical Distinctions):

Misunderstanding 22:
"Consciousness is a form of information processing."
Clarification:

Information processing is syntactic; consciousness is qualitative and experiential.

Misunderstanding 23:
"Quantum mechanics creates consciousness."
Clarification:
Quantum processes lack intentionality, subjectivity, and qualitative character.

Misunderstanding 24:
"Consciousness is reducible to brain states or computations."
Clarification:
Reductionist accounts fail to explain qualia, intentionality, and unified subjectivity.

4.5 Misunderstandings Concerning Theology (Questions Most Readers Have):

Misunderstanding 25:
"God is part of the universe or a force in nature."
Clarification:
Classical theism maintains that God is transcendent, simple, uncreated, and not part of the physical cosmos.

Misunderstanding 26:
"Creation refers to a temporal beginning."
Clarification:
Creation refers to ontological dependence, not necessarily a temporal starting point.

Misunderstanding 27:
"Providence means deterministic control of events."
Clarification:
Providence includes primary causation without negating secondary causes or human freedom.

Misunderstanding 28:
"Divine action competes with natural causes."
Clarification:
Primary and secondary causation operate on different levels; they do not compete.

Misunderstanding 29:

"Imago Dei means physical resemblance or superior intelligence."

Clarification:

Image-bearing refers to rationality, relationality, freedom, and spiritual capacity.

(Advanced Technical Distinctions):

Misunderstanding 30:

"The Logos is equivalent to mathematical structure or physical order."

Clarification:

Logos includes rationality and order but is in essence personal and divine, not merely structural.

SECTION 5
ANTICIPATED OBJECTIONS AND SCHOLARLY RESPONSES

This section addresses technical objections from academic specialists. You can skip this entire section without missing anything essential to understanding the synthesis. It's included for scholars, educators, and readers who want to engage with specialized critiques.

If you're reading this book to understand how faith and science fit together, Sections 1-4 and 6-7 provide everything you need. Come back to Section 5 later if you're curious about how the framework responds to scholarly criticism.

5.1 Objections from Physics
Objection 1:
"This framework misuses terms like 'information,' 'order,' or 'observer' from physics."

Response:
In physics, these terms have strictly defined operational meanings. This work employs them only within their scientific domains and does not project them into metaphysical or theological discourse.

When used analogically, the analogical nature is explicitly

signaled, and no physical inference is drawn from metaphysical premises.

Objection 2:

"It conflates quantum potentiality with metaphysical potentiality."

Response:

Quantum potentiality refers to probability amplitudes within Hilbert space. Metaphysical potency refers to the ontological capacity of beings to become otherwise. These belong to distinct explanatory frameworks and are not equated. Any resemblance in terminology is acknowledged as linguistic coincidence, not conceptual overlap.

Objection 3:

"It treats the holographic principle as ontological rather than mathematical."

Response:

The holographic principle is treated as a formal structural correspondence within theoretical physics.

No ontological conclusions are drawn from holography regarding the nature of existence itself.

Its use is carefully confined to discussing structural parallels, not metaphysical claims.

Objection 4:

"It implies that consciousness affects quantum measurement."

Response:

Nothing in this work relies on consciousness-induced collapse. Measurement is consistently treated as a physical interaction, in line with mainstream interpretations of quantum theory and ongoing experimental verification.

Objection 5:

"It blurs physical cause with metaphysical cause."

Response:

Physical causation concerns dynamical relations governed by equations. Metaphysical causation concerns dependence relations underlying existence itself. These categories do not compete or overlap.

5.2 Objections from Philosophy of Mind

Objection 6:

"It assumes dualism without establishing it."

Response:

No commitment is made to any specific dualist model. The irreducibility of consciousness is treated as a phenomenological and philosophical fact, not as a metaphysical dogma. The analysis preserves conceptual distinctions without requiring a full ontological account.

Objection 7:

"It treats consciousness as fundamental without argument."

Response:

Consciousness is treated as irreducible, not as a fundamental metaphysical primitive.

The argument is that no physical, functional, or computational theory currently accounts for subjectivity. This position aligns with mainstream analytic philosophy of mind.

Objection 8:

"It confuses intentionality with information processing."

Response:

Intentionality is recognized as a property of mental states, not algorithms or data structures.

Information theory deals with syntactic distinctions, not semantic directedness. These categories remain distinct throughout.

Objection 9:

"It risks sliding into panpsychism."

Response:

The framework intentionally avoids
panpsychism by:
distinguishing structural relations from conscious awareness
grounding interiority uniquely in rational subjects

denying consciousness to non-personal or non-rational systems
No attribution of subjective awareness is made to matter or energy.

5..3 Objections from Metaphysics

Objection 10:

"It conflates physical contingency with metaphysical contingency."

Response:

Physical contingency refers to changes within physical states or processes. Metaphysical contingency concerns the dependence of a being on a cause of existence. These distinctions are maintained rigorously.

Objection 11:

"It overuses teleological language in contexts where finality is inappropriate."

Response:

Teleology is employed strictly in its metaphysical sense: intrinsic directedness within being itself. No teleology is imposed upon physical models except where classical metaphysics properly interprets final causality. Any discussion of purpose or directionality in the universe is strictly metaphysical or theological in scope, not a physical mechanism.

Objection 12:

"It risks treating metaphysical necessity as if it were scientifically demonstrable."

Response:

Metaphysical necessity is grounded in ontological reasoning, not empirical demonstration.

This work maintains the distinction between logical, metaphysical, and physical necessity.

5.4 Objections from Theology

Objection 13:

"It collapses God into abstract metaphysical categories."

Response:

Classical theism affirms that God is simple, infinite, and transcendent. Metaphysical description is consistent with theological revelation and does not reduce God to abstraction.

Objection 14:

"It diminishes divine transcendence by discussing structural order in the universe."

Response:

Order and intelligibility in creation point to coherence but do not constrain or define the divine nature. Transcendence is upheld fully; creation remains ontologically distinct.

Objection 15:

"It risks suggesting God is identical with information, order, or rational structure."

Response:

God is not equated with information or structure. These are properties of creation; God is the transcendent ground of being and intelligibility.

Objection 16:

"It implies God must act through natural laws as mechanisms."

Response:

Primary causation and secondary causation are distinct. God's action is ontological, not mechanical; natural laws are descriptions of creaturely behavior.

5.5 Objections from Skeptics and Naturalists

Objection 17:

"It smuggles theological conclusions into physical or informational language."

Response:

No physical argument is presented as proving metaphysical or theological claims. The disciplines are kept rigorously separate; analogies are explicitly marked.

Objection 18:

"It assumes meaning exists objectively."

Response:

Meaning is grounded in rational subjects and is recognized as a real feature of mental life.

This position is consistent with mainstream philosophy and does not rely on supernatural assumptions.

Objection 19:
"It assumes purpose in the universe."
Response:
Teleological language is metaphysical, not empirical. Purpose is not imposed upon physical processes but analyzed within a classical metaphysical framework.

Objection 20:
"It arbitrarily prefers classical metaphysics over scientific naturalism."
Response:
Classical metaphysics addresses questions naturalism cannot answer:

- existence
- identity
- explanation
- intelligibility
- consciousness
- rational order

The two are not competitors but distinct modes of inquiry.

5.6 Integrated Summary of Responses

- Physics describes mechanisms; metaphysics describes being.
- Quantum theory provides structural models, not ontological commitments.
- Information theory measures distinctions, not meaning or consciousness.
- Consciousness is irreducible and cannot be captured by physical explanation.
- Theology addresses the personal, transcendent source of all being.

- No discipline collapses into another.
- Analogical language is not literal description.
- Structural parallels are not metaphysical identities.
- Divine action does not compete with natural causality.
- Interpretive arguments are clearly separated from empirical claims.

SECTION 6
CROSS-DISCIPLINE TRANSLATION TABLES

Cross-disciplinary work often involves words that appear identical but function differently depending on the conceptual framework.

6.1 "Information" Across Domains

Domain	Meaning of 'Information'	Nature	Notes
Physics	Physical states and distinctions; measurable correlations	Quantitative	No meaning or intentionality
Information Theory	Shannon information; uncertainty reduction	Mathematical	Purely syntactic
Computer Science	Encoded symbols for processing	Formal	Depends on rules and algorithms
Biology	Genetic sequences encoding functions	Functional	Not semantic; not metaphysical
Philosophy of Mind	Content of mental representations	Semantic	Requires a subject
Theology	Divine knowledge (not "information")	Personal	Not reducible to data

Key Insight: Information is never equated across these domains; each uses the term in a distinct, irreducible sense.

6.2 "Cause" Across Domains

Domain	Meaning of "Cause"	Explanation Type
Physics	Forces, interactions, dynamical laws	Mechanistic
Metaphysics	Act by which a being exists or becomes	Ontological
Theology	Divine sustaining of existence (primary cause)	Transcendent
Biology	Functional or evolutionary pressures	Teleofunctional
Information Theory	Constraints affecting signal or message	Structural
Philosophy of Mind	Intentional or rational agency	Personal/Mental

Key Insight: A cause in one domain cannot be substituted for a cause in another.

6.3 "Order" Across Domains

Domain	Meaning of "Order"	Mode
Physics	Symmetry, invariance, and structure	Mathematical
Metaphysics	Finality and intrinsic directedness	Teleological
Theology	Providence and divine wisdom	Personal
Information Theory	Structured patterns in data	Formal
Biology	Organization enabling life	Functional

Key Insight: Order is not a single concept — mathematical, teleological, personal, and formal orders are distinct.

6.4 "Observer" Across Domains

Domain	Meaning	Involves Consciousness?
Physics (QM)	Any interacting system that produces definite outcomes	No
Relativity	A reference frame for measurement	No
Information Theory	Receiver or decoder of signals	Not necessarily
Philosophy of Mind	Conscious subject experiencing reality	Yes
Theology	God as the knower of all things	Yes, personal

Key Insight: "Observer" in physics is never a mind; it becomes personal only in philosophy of mind or theology.

6.5 "Potential" Across Domains

Domain	Meaning	Involves Consciousness?
Physics (QM)	Any interacting system that produces definite outcomes	No
Relativity	A reference frame for measurement	No
Information Theory	Receiver or decoder of signals	Not necessarily
Philosophy of Mind	Conscious subject experiencing reality	Yes
Theology	God as the knower of all things	Yes, personal

Key Insight: Quantum potential ≠ metaphysical potency.

6.6 "Law" Across Domains

Domain	Meaning	Ontological Status
Physics	Mathematical description of behavior	Descriptive, not prescriptive
Metaphysics	Principle of being	Prescriptive in structure
Theology	Divine moral or providential order	Personal
Logic	Rules governing valid inference	Formal
Ethics	Norms of right action	Normative

Key Insight: Laws of nature describe; divine law commands; logical laws structure thought.

6.7 "Word / Logos / Reason" Across Domains

Domain	Term	Meaning
Greek Philosophy	Logos	Rational principle of order
Hebrew Thought	Davar	Active, creative speech
Christian Theology	Logos	The eternal Word, personal
Linguistics	Word	Symbolic linguistic unit
Information Theory	Codeword	Symbol sequence in a message

Key Insight: Logos is personal and theological; not merely structural or linguistic.

6.8 "Knowledge" Across Domains

Domain	Meaning	Source
Human Cognition	Mental grasp of propositions	Learning, experience
AI / Computation	Stored data or weights	Algorithms
Biology	Encoded survival-related behavior	Evolution
Theology	God's perfect knowledge of all things	Divine essence

Key Insight: Divine knowledge is not computational or acquired.

6.9 "Being" Across Domains

Domain	Meaning	Notes
Physics	Physical entity with measurable properties	Exists in spacetime
Metaphysics	Anything that exists in any mode	Includes immaterial realities
Ontology	Category of existence	Formal structure
Theology	God as *ipsum esse subsistens*	Not a being among beings

Key Insight: Being is analogical — physics deals with one mode; metaphysics with all modes; theology with the source.

6.10 "Spirit / Breath / Life" Across Domains

Domain	Meaning	Clarification
Hebrew Thought	Ruach	Divine presence and power
Biology	Life processes	Cellular and biochemical activity
Physics	Not defined	No equivalent concept
Theology	Holy Spirit (personal)	Not an energy or force

Key Insight: Spirit is personal; biology and physics do not recognize a comparable category.

6.11 "Order of Explanation" Table

Question Type	Addressed By
Mechanism	Physics
Existence	Metaphysics
Purpose	Metaphysics / Theology
Personal meaning	Philosophy of Mind / Theology
Structure & signals	Information Theory
Conscious experience	Philosophy of Mind
Moral value	Theology / Ethics

Key Insight: Discipline-specific limits must be honored for coherence.

SECTION 7
INTEGRATION GUIDELINES

This section establishes the explicit methodological rules for combining insights across different fields while preserving their boundaries. The goal is not to merge disciplines into one system, but to prevent confusion by ensuring that each field contributes only what properly belongs to it.

7.1 The Principle of Complementarity

Different disciplines offer complementary, not competing, modes of explanation.

- Physics explains mechanisms.
- Metaphysics explains being and causation.
- Information theory explains structure and distinctions.
- Philosophy of mind explains subjectivity and mental states.
- Theology explains ultimate origin, purpose, and divine action.

No discipline cancels the others; each operates on its proper level.

7.2 The Principle of Non-Substitution

No discipline may substitute for another. Examples:

- Physical laws cannot substitute for metaphysical explanation.
- Information theory cannot substitute for consciousness.
- Quantum mechanics cannot substitute for theology.
- Metaphysics cannot substitute for empirical science.
- Theology cannot substitute for physics or cosmology.

Each provides answers appropriate to its own domain.

7.3 The Principle of Hierarchical Explanation

Reality contains layers of explanation, each appropriate to a different domain:

- Physical layer — behavior, mechanisms, processes
- Biological layer — life, function, adaptation
- Psychological layer — experience, awareness, intention
- Metaphysical layer — being, identity, causality
- Theological layer — Creator, meaning, purpose

Higher-level explanations do not override lower-level mechanisms; lower-level explanations do not eliminate higher-level meaning.

7.4 The Principle of Non-Reduction

This book suggests and defends the view that complex phenomena cannot be reduced to a single explanatory framework. This suggests that:

- Consciousness cannot be exhaustively reduced to computation or neural activity.
- Meaning is not reducible to information theory.
- Providence is not reducible to physical laws.
- Existence is not reducible to physics.
- Teleology is not reducible to biological function.

Reduction erases essential aspects of reality.

7.5 The Principle of Analogical Interpretation

When terms are used across domains, they must be recognized as analogical, not univocal.

Examples:

- "Word" (Logos) ≠ linguistic symbol
- "Light" (theological) ≠ electromagnetic radiation
- "Spirit" ≠ physical energy
- "Order" (metaphysical) ≠ mathematical symmetry
- "Cause" (metaphysical) ≠ force (physics)

Analogical bridges must be explicitly marked as analogical.

7.6 The Principle of Conceptual Integrity

A concept must retain its meaning from its home discipline even when used metaphorically.

For example:

- "Information" retains its syntactic meaning unless clearly reinterpreted.
- "Observer" retains its operational meaning in physics.
- "Being" retains its metaphysical meaning.
- "Logos" retains its theological meaning.

No concept may silently shift domains without clarification.

7.7 The Principle of Structural Resonance

Permissible Integration:

Disciplines may illuminate each other through parallels or resonances, as long as:

- the structural analogy is explicitly stated
- the analogy is not taken as identity
- the domains remain distinct

Examples of permissible resonance:

- Symmetry in physics ↔ rational order in metaphysics

- Intelligibility of the universe ↔ Logos theology
- Information-theoretic structure ↔ metaphysical form
- Unity of consciousness ↔ metaphysical simplicity (at the creaturely level)

These comparisons are conceptual bridges, not assertions of identity.

7.8 The Principle of Interpretive Modesty

No interdisciplinary claim should exceed the limits of its contributing domains.

For example:

- A physical model cannot settle metaphysical debates.
- A metaphysical argument cannot predict scientific results.
- A theological doctrine cannot dictate empirical findings.
- A theory of consciousness cannot explain cosmological behavior.

Each field contributes insight, but none oversteps its competence.

7.9 The Principle of Dual Explanation (Non-Competition)

When something has multiple dimensions, its explanations may coexist without conflict.

Example:

- A written letter can be explained by:
- Physics: ink patterns on paper
- Information theory: encoded symbols
- Linguistics: grammatical structure
- Philosophy of mind: intention and meaning
- Theology/metaphysics: rational agent's existence and purpose

These explanations address different questions about the same thing without contradiction.

7.10 The Principle of Ontological Priority

Metaphysics and theology provide the most fundamental explanations, but do not negate physical explanation.

- Physics explains how things behave.
- Metaphysics explains why things exist and what they are.
- Theology explains who grounds all being and order.

Ontological priority does not imply empirical or methodological priority.

7.11 The Principle of Interdisciplinary Honesty

Any time a term is borrowed across domains, this work explicitly:

- defines it
- situates it
- marks analogy vs. literal usage
- clarifies what is not implied

This ensures no reader mistakes metaphor, parallel, or resonance for identity.

SECTION 8
COMPARATIVE FRAMEWORKS: CLARIFYING THE (IAM) BY CONTRAST

This section situates the (IAM) within the broader landscape of contemporary metaphysical and cosmological frameworks that address similar foundational questions concerning existence, information, consciousness, and order. Several modern theories employ concepts such as self-reference, information, rational structure, and emergent meaning in ways that may appear, at first glance, to overlap with elements of this work.

The purpose of this comparison is not polemical. It is clarificatory. By examining points of apparent similarity alongside points of fundamental divergence, this section aims to prevent category errors, misclassification, and unintended philosophical conflations.

Each framework discussed below operates under distinct ontological assumptions, methodological commitments, and explanatory goals. Understanding these differences is essential for correctly interpreting what the (IAM) claims, and what it explicitly does not claim.

8.1 The Cognitive-Theoretic Model of the Universe (CTMU)
Christopher Michael Langan
Primary Sources:

Langan, C. M. The Cognitive-Theoretic Model of the Universe (various unpublished manuscripts and essays, 1989–present).

Langan, C. M. "The (CTMU): A New Kind of Reality Theory." (Online essays and interviews).

Core Claim:

The Cognitive-Theoretic Model of the Universe (CTMU) proposes that reality is a Self-Configuring, Self-Processing Language (SCSPL). According to this framework, reality is both syntactic (law-like), semantic (meaning-bearing), and pragmatic (self-selecting). The universe is said to explain its own existence through intrinsic self-reference, thereby eliminating the need for any ontologically external ground.

Within the (CTMU), God is identified with the total self-referential structure of reality itself. Existence is logically necessary, and reality is conceived as a closed explanatory system that contains both being and explanation within a single formal structure.

Points of Apparent Overlap:

At a conceptual level, several themes in the (CTMU) may appear similar to those explored in the (IAM):

- Emphasis on self-reference as unavoidable in any complete theory of reality
- Recognition that reality is intelligible and structured by rational constraints
- Use of information and language metaphors to describe physical law
- Rejection of reductive materialism
- Affirmation that consciousness is not an accidental byproduct of matter

These surface similarities often lead readers to assume a close philosophical alignment between the two frameworks.

Key Divergences:

Despite these superficial resemblances, the (CTMU) and the (IAM) diverge at the most fundamental ontological level.

Ontology vs. Description

In the (CTMU), language is ontological. Reality literally is a self-processing linguistic structure.

In the (IAM), information and language are descriptive tools, not ontological substrates. They describe how physical systems behave without constituting being itself.

Creator–Creation Distinction

The (CTMU) explicitly collapses the distinction between God and reality; God is identified with the totality of existence.

The (IAM) preserves the classical metaphysical distinction between Creator and creation. God is Being Itself; the universe participates in being but is not identical with its source.

Necessity vs. Freedom

In the (CTMU), reality exists by logical necessity. Creation is unavoidable.

In the (IAM), God exists necessarily, but creation is free, contingent, and not logically compelled.

Personhood and Relationality

The (CTMU) describes a form of abstract, system-level self-awareness, but not a personal God capable of genuine relationality.

The (IAM) affirms a personal, relational God, fully realized in Trinitarian theology and historically revealed through the Logos.

Why the Difference Matters:

The (CTMU) offers a logically ambitious and internally coherent framework, but it does so by identifying explanation with necessity and collapsing metaphysical distinctions that classical philosophy has long regarded as essential. In doing so, it eliminates genuine contingency, freedom, and relationality.

The (IAM) rejects the idea that ultimate explanation must take the form of logical closure. Instead, it affirms that reality is intelligible precisely because it is grounded in a rational, personal source whose freedom is not reducible to formal necessity. This distinction preserves moral responsibility, meaningful love, and the coherence of historical revelation.

8.2 Digital Physics and "It from Bit"

John Archibald Wheeler and Related Approaches
Primary Sources:
Wheeler, J. A. "Information, Physics, Quantum: The Search for Links." Proceedings of the 3rd International Symposium on Foundations of Quantum Mechanics (1989).

Fredkin, E. "Digital Mechanics." Physica D 45 (1990).

Lloyd, S. Programming the Universe (2006).

Core Claim:
Digital physics proposes that the universe is fundamentally computational or informational in nature. Physical reality is described as emerging from discrete informational processes, often analogized to computation or digital state transitions.

Key Divergence:
The (IAM) explicitly rejects ontological computationalism. While information theory is employed descriptively to understand physical law, the universe is not treated as a computer, simulation, or algorithmic substrate. Information is physical and constrained by matter and energy; it is not the ground of being.

8.3 Panpsychism and Cosmopsychism
Representative Thinkers:
Galen Strawson, Philip Goff, Bernardo Kastrup

Primary Sources:
Strawson, G. "Realistic Monism." Journal of Consciousness Studies (2006).Goff, P. Galileo's Error (2019).

Kastrup, B. The Idea of the World (2019).

Core Claim:
These frameworks propose that consciousness is a fundamental feature of reality, either present in all matter (panpsychism) or instantiated as a universal cosmic mind (cosmopsychism).

Key Divergence:
The (IAM) does not posit consciousness as a universal intrinsic property of matter. Consciousness arises within created persons and reflects, rather than constitutes, the divine nature. God is conscious, but creation is not God.

8.4 Classical Deism and Methodological Naturalism

Core Claim:

Deism posits a non-interventionist creator, while methodological naturalism brackets metaphysical questions entirely, restricting inquiry to empirical mechanisms.

Key Divergence:

The (IAM) rejects both positions by affirming that metaphysical grounding and historical revelation are not violations of scientific integrity but operate in complementary explanatory domains.

8.5 Summary Comparison Tables

Framework	Ground of Reality	God-World Relation
CTMU (Langan)	Self-referential logic	Identity
Digital Physics	Computation	Undefined
Panpsychism	Consciousness	Diffuse
Deism	Remote creator	Detached
(IAM)	Being Itself "I AM"	Creator–creation distinction

Table 1: Ontological Fondations

Framework	Role of Information	Freedom	Personhood
CTMU (Langan)	Ontological	No	Abstract
Digital Physics	Ontological	Limited	None
Panpsychism	Secondary	Ambiguous	Impersonal
Deism	Minimal	Yes	Weak
(IAM)	Descriptive	Yes	Personal

Table 2: Information, Freedom, Personhood

Concluding Clarification:

The (IAM) does not compete with these frameworks on their

own terms. It operates from a different starting point: the classical metaphysical claim that being itself requires grounding beyond formal structure, mechanism, or self-reference. Information, logic, and law describe reality; they do not generate it. The universe is intelligible not because it explains itself, but because it proceeds from the Logos—the eternal 'I AM' through whom all things were made.

TWO PRAYERS OFFERED WITH RESPECT FOR PROTESTANT AND EVANGELICAL PRACTICE

RESPONSES AND PRAYERS

The following prayers are examples, not formulas. Christian traditions differ in how they normally express repentance, confession, and commitment. If you are Catholic or Orthodox, the ordinary path includes life within the Church and its sacramental practices. If you are Protestant, you may express this as personal repentance and trust in Christ. Use the form that fits your tradition or simply speak honestly to God.

TWO PRAYERS OFFERED WITH RESPECT TO CATHOLIC TEACHING AND PRACTICE

A Prayer of Repentance and Return
Father in heaven, I turn to You.
I confess that I have sinned in thought, word, and deed, and that I have failed to love You and others as I should. I am sorry for my sins, and I desire to change.
Lord Jesus Christ, Son of the living God, have mercy on me, a sinner. I trust in Your Cross and Resurrection. By Your grace, bring me back to friendship with You.
Holy Spirit, give me a contrite heart, the courage to seek reconciliation, and the desire to live faithfully. Lead me into the life of the Church. Help me to pray, to learn the faith, and to walk in love.
Mary, Mother of Jesus, pray for me; and all you saints, pray for me, that I may follow Christ with a whole heart.
Amen.

Prayer for Faith and Understanding
Heavenly Father,
I thank You for the gift of existence, for the order of creation, and for the light of reason.
Give me humility when I study, patience when I struggle, and honesty when I doubt.
Lord Jesus Christ, Son of the living God
draw me closer to You.
Teach me to trust Your Church, to love what is true, and to follow where You lead.
Holy Spirit,
illumine my mind and strengthen my will.
Help me to seek wisdom, to love my neighbor,
and to live with reverence for God and for every human person.
If I have wandered, bring me home.
If I am afraid, give me courage.
If I am proud, give me humility.
Mary, Mother of God, pray for me.
All holy angels and saints, pray for me.
Amen.

TWO PRAYERS OFFERED WITH RESPECT TO EASTERN ORTHODOX TEACHING AND PRACTICE

A Prayer for Mercy and Communion

Lord Jesus Christ, Son of God,
have mercy on me, a sinner.

I come to You with my weakness and my wandering. I have not loved You with my whole heart. I have wounded myself and others by sin. I do not justify myself. I ask for Your mercy.

Cleanse me. Heal me. Restore me. Unite my heart to fear Your name and to love You. Grant me repentance, humility, and patience.

Holy Spirit, breathe life into me. Teach me to pray, to forgive, and to seek peace. Bring me into the life of Your Church, that I may learn to worship, to repent, and to grow in the likeness of Christ.

Into Your hands I place my life—today and always.

Amen.

Prayer for Illumination and Mercy
O Lord Jesus Christ, Son of God,
have mercy on me and guide me into truth.
Most Holy Trinity,
Father, Son, and Holy Spirit,
grant me purity of heart and clarity of mind.
Keep me from pride, from vanity, and from using knowledge
without love.
Teach me to see creation with reverence,
to honor the image of God in every person,
and to seek wisdom as a form of obedience.
Illuminate my understanding,
heal what is broken in me,
and make my life a humble offering to You.
For You are good and You love mankind.
Amen.

TWO PRAYERS OFFERED WITH RESPECT FOR PROTESTANT AND EVANGELICAL PRACTICE

Sinner's Prayer of Salvation

Heavenly Father,

I come to you now a sinner,

I believe Jesus Christ is Lord. I believe You sent Him for us, that He died for our sins, and that You raised Him from the dead.

Lord Jesus, have mercy on me. Forgive me. Save me. I turn from sin, and I turn to You. Teach me to follow You, to obey You, and to love what You love.

Holy Spirit, make me new. Give me a new heart and a steady faith. Lead me into Your truth, into Your Church, and into a life that honors You.

I receive Your grace with gratitude, and I place my trust in Christ alone.

Amen.

Prayer of Trust in Christ
Heavenly Father,
I come to You with an open heart. I need Your mercy, Your forgiveness, and Your guidance.
Jesus Christ,
I believe You are Lord.
I believe You died for my sins and rose again.
I turn from my sin and place my trust in You.
Please forgive me, change me, and lead me.
Holy Spirit,
give me new life and a steady faith.
Teach me to love God and love others.
Give me courage to follow Christ, even when it costs me something.
Thank You for saving me by grace, not by my works.
I belong to You.
Amen.

NEXT STEPS

If you prayed one of these prayers or perhaps felt called to God, the next steps are for you. They are about learning the faith, receiving grace, and becoming grounded in a real community.

Join a local church community:

- Don't try to do faith alone. Find a church where Scripture is taken seriously, Christ is central, and love is practiced.
- If you're unsure where to start, visit a few: a Catholic parish, an Orthodox parish, and a Protestant church with a clear commitment to historic Christianity. Pay attention to worship, teaching, and the character of the community.

Speak with a pastor or priest:

- Tell them you're exploring Christianity, and you want a grounded path forward.
- Ask simple questions: "What do you believe?" "How do people grow?" "What would you recommend I do next?"

Begin with the Gospels:
Start reading Luke or John, slowly. Don't rush.
A simple approach. Read a short section, ask:

- "What does this show me about Jesus?"
- "What would it look like to live this?"

Make prayer part of your life!
Learn the basics without drowning in it:
You don't need a PhD. You need a foundation.
Ask your church for an intro path:

- Catholic: RCIA / OCIA (a guided introduction to the faith)
- Orthodox: an inquirer/catechumen path
- Protestant: a basics class, small group, or discipleship course

About baptism, communion, and confession (why traditions differ):

- All historic Christian traditions treat baptism as a major beginning—entry into the life of the Church in some form.
- Catholic and Orthodox Christians normally understand grace as something God gives through the life of the Church, including sacraments, and will invite you into a structured path.
- Protestant Christians emphasize personal repentance and trust in Christ and will also encourage baptism and ongoing discipleship as obedience to Christ.
- If you're uncertain, start with Jesus: keep reading the Gospels, stay in community, and ask wise leaders to help you discern.

Keep the posture that keeps you sane:

- Be honest. Be patient. Don't pretend.
- Faith is not anti-science. It is not anti-reason. It is a commitment of trust grounded in truth, lived out over time.
- If something feels confusing, don't run from questions—bring them into conversation with thoughtful believers who respect both faith and careful thinking.

APPENDIX B

ETYMOLOGY

This appendix presents key biblical words in their original Hebrew or Greek forms, along with their meanings and theological significance. These terms are central to understanding how Scripture describes creation, revelation, and the nature of God. They also demonstrate how the biblical worldview aligns with the informational and relational structure of the universe.

HEBREW AND GREEK WORD STUDIES

Davar – דָּבָר – "dah-var" {Meaning: word, speech, matter, event, action} Davar does not refer merely to spoken language. It is a word that accomplishes something. In Genesis, God creates by speaking. His words bring reality into existence. Davar represents a meaningful act that establishes order and structure. In an informational universe, davar expresses the idea that creation arises from meaningful articulation rather than random physical processes. *Key References: Genesis 1:3, Psalm 33:6, Isaiah 55:11.*

Chokhmah – חָכְמָה – "khok-mah" {Meaning: wisdom, skill, order, structure} Chokhmah includes practical skill, moral discernment, and the underlying structure of creation. *Proverbs 8* describes wisdom as present with God before creation, delighting in His work. Wisdom is not merely intellectual. It is the deep patterning of the universe. In modern language, chokhmah corresponds to the informational order and natural law embedded in the universe. *Key References: Proverbs 3:19, Proverbs 8:22–31, Job 28.*

Ruach – רוּחַ – "roo-akh" {Meaning: breath, wind, spirit} Ruach conveys movement, life, energy, and presence. In *Genesis 1*, the

Spirit of God moves over the waters before creation takes shape. Ruach is the animating, sustaining breath of God in creation. In the (IAM), ruach corresponds to God's continual sustaining activity within the universe. *Key References: Genesis 1:2, Ezekiel 37:1–14, Psalm 104:30.*

Ehyeh Asher Ehyeh – אֶהְיֶה אֲשֶׁר אֶהְיֶה – "eh-yeh asher eh-yeh" {Meaning: "I AM That I AM"} This is the name God reveals to Moses at the burning bush. It expresses God's self-existence, eternal presence, and unchanging nature. God is not one being among others. He is the ground of all being, the One who simply is. This name is foundational for the (IAM), which understands God as the source of all existence. *Key References: Exodus 3:14, Isaiah 41:4, Revelation 1:8.*

Logos – Λόγος – "loh-gos" {Meaning: word, reason, order, rational structure} Logos is a central term in Greek philosophy and the New Testament. It can refer to speech, reason, or an organizing principle. The Stoics used it to describe the rational structure permeating the universe. John's Gospel applies this term to Jesus Christ, declaring that the Logos through whom all things were made became flesh. This connects the rational, relational structure of the universe with the person of Christ. *Key References: John 1:1–14, Colossians 1:15–17, Hebrews 1:1–3.*

Pneuma – Πνεῦμα – "pnyoo-mah" {Meaning: spirit, breath, wind, life} Pneuma is the Greek counterpart to ruach. It refers to the Holy Spirit, the breath of life, and the inner vitality of living beings. In the New Testament, the Spirit indwells believers, gives life, and empowers them for service. Pneuma represents God's active presence, binding together the relational order of creation. *Key References: John 3:5–8, Romans 8:9–16, 1 Corinthians 12:4–13.*

Doxa – Δόξα – "dox-ah" {Meaning: glory, radiance, weight, honor} Doxa refers to the visible manifestation of God's presence. It is not mere brightness but the expression of His nature. In biblical thought, glory is the outward display of God's beauty, holiness, and majesty. In the (IAM) framework, doxa reflects the visible expression of God's presence within the relational universe. *Key References: Exodus 33:18–19, John 17:5, Revelation 21:23.*

APPENDIX C

DIAGRAMS AND ILLUSTRATIONS

DIAGRAMS AND ILLUSTRATIONS

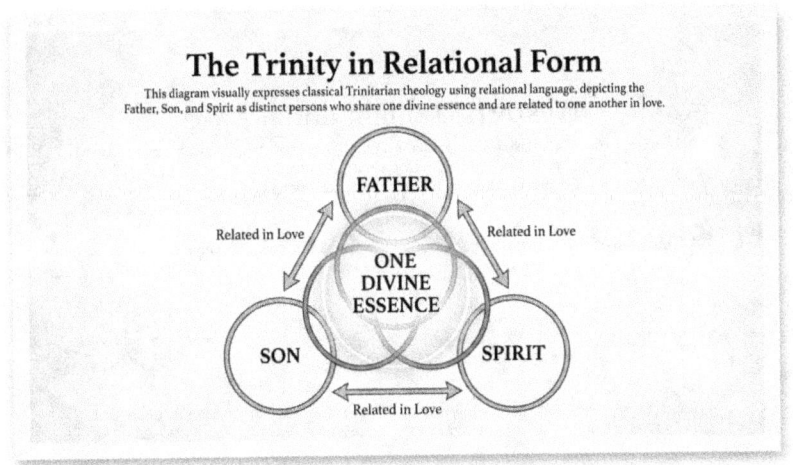

Trinity in Relational Form

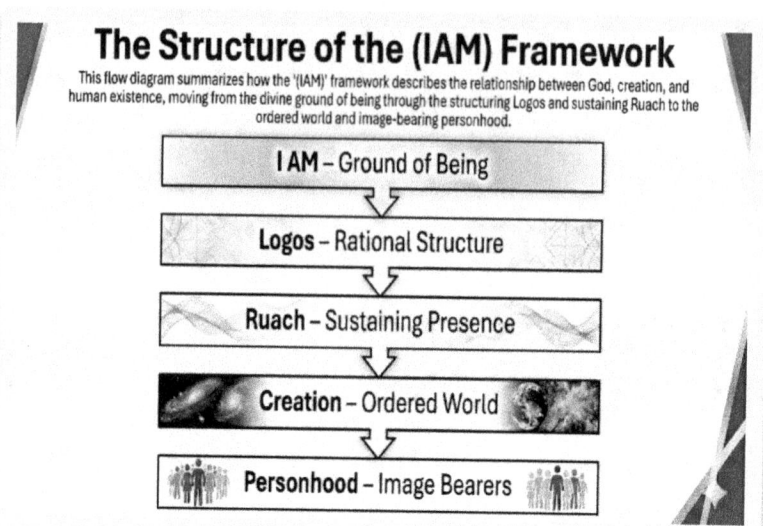

Structure of the (IAM) Framework

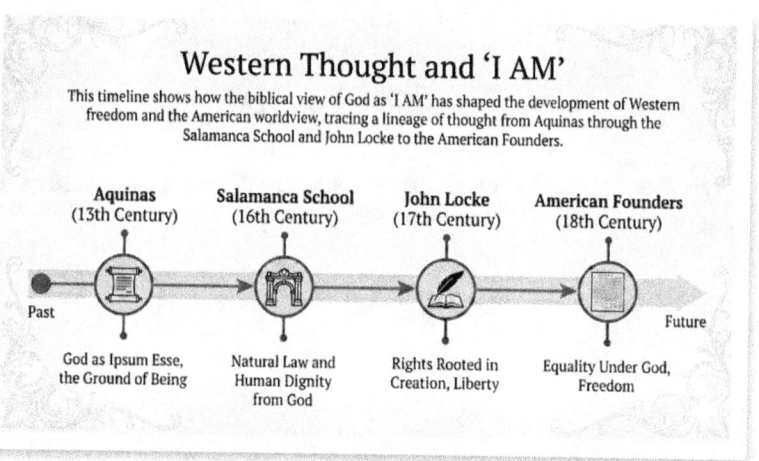

Western Thought and 'I AM'

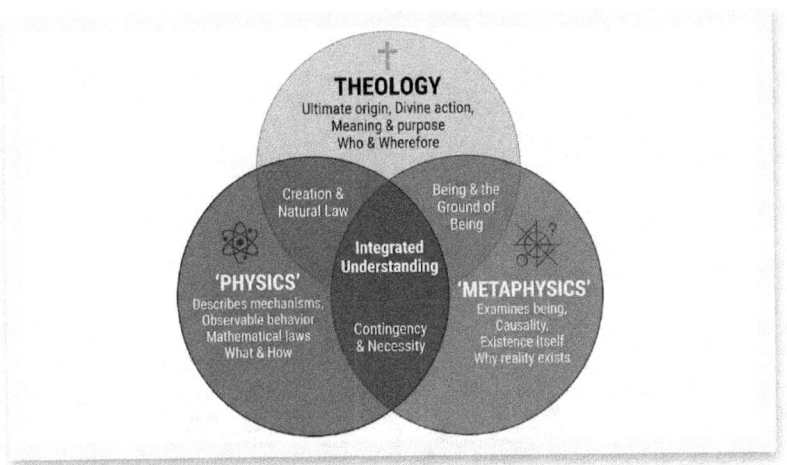

The Three Domains of Inquiry: Each discipline addresses different questions using distinct methodologies. Respecting these boundaries enables genuine interdisciplinary dialogue without methodological confusion. (See Part X, Sections 1-2)

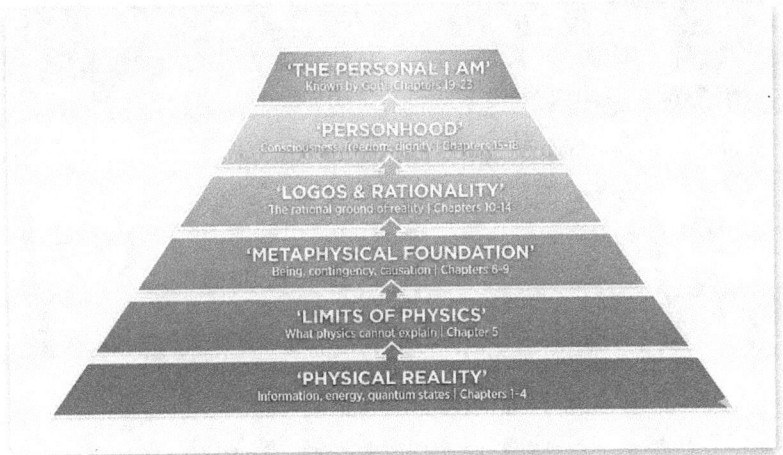

The Argumentative Structure: From the physics of information to the personal invitation of I AM. Each level builds upon and requires the insights of the previous levels. (See Introduction: How to Use This Book)

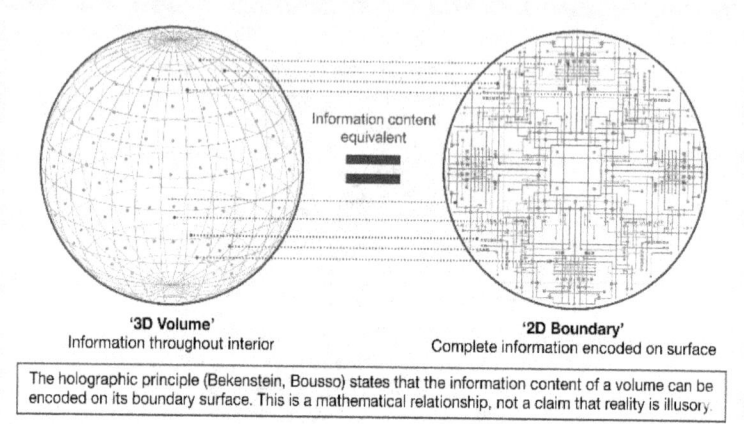

The Holographic Principle: A mathematical framework showing the deep relationship between information, geometry, and physical reality. (Chapter 3, Part X Section 4)

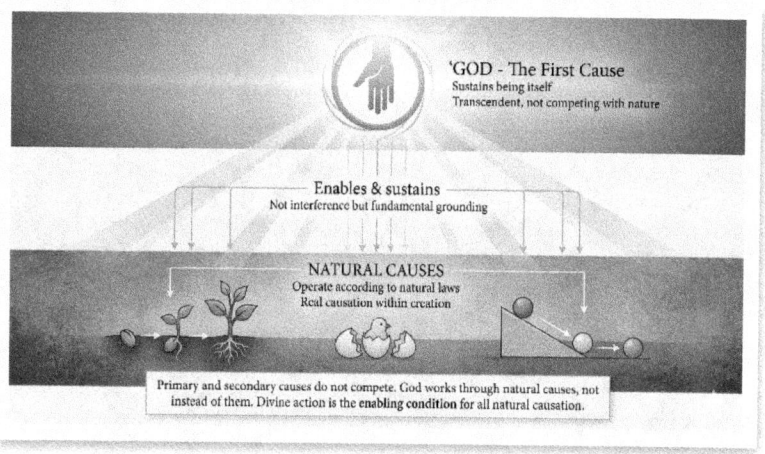

Primary and Secondary Causation: Understanding how divine and natural causes relate without competing. God's action is the ground of nature's action, not an intervention into it. (Chapters 10-11, Part X)

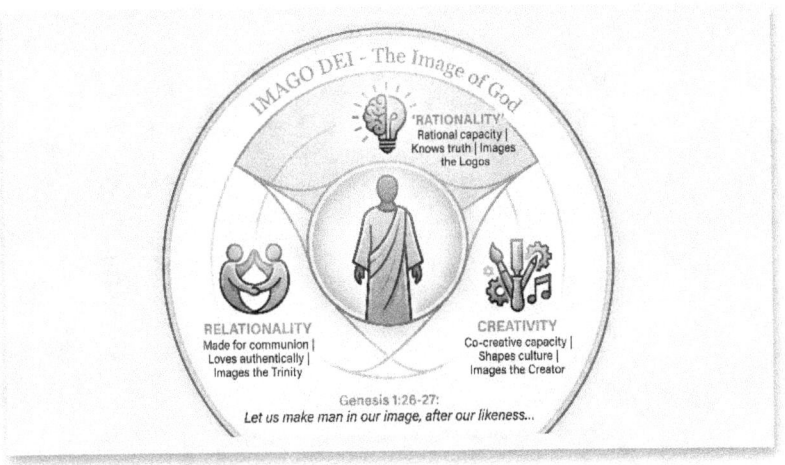

The Image of God: Human persons bear God's image through rationality (knowing), relationality (loving), and creativity (making). We image the Triune God who is Logos, Love, and Creator. (Chapters 15-18, Part VI)

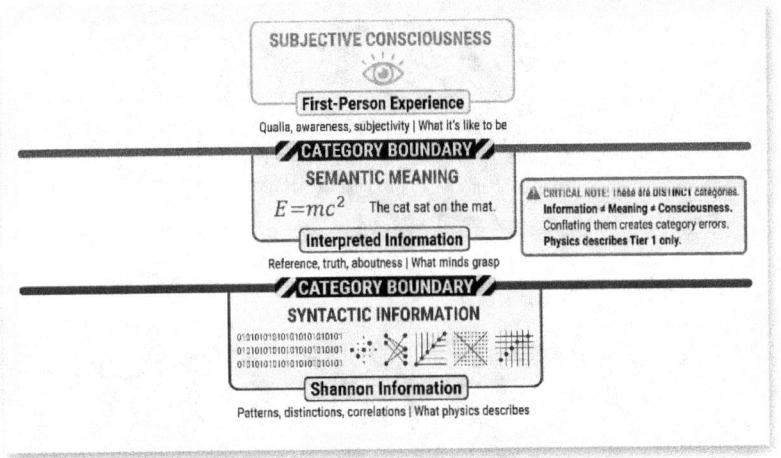

Three Distinct Categories: Information (syntactic patterns), Meaning (semantic content), and Consciousness (subjective experience) are fundamentally different. Respecting these distinctions prevents category errors. (Ch 7, Part X Sections 2, 4, 7)

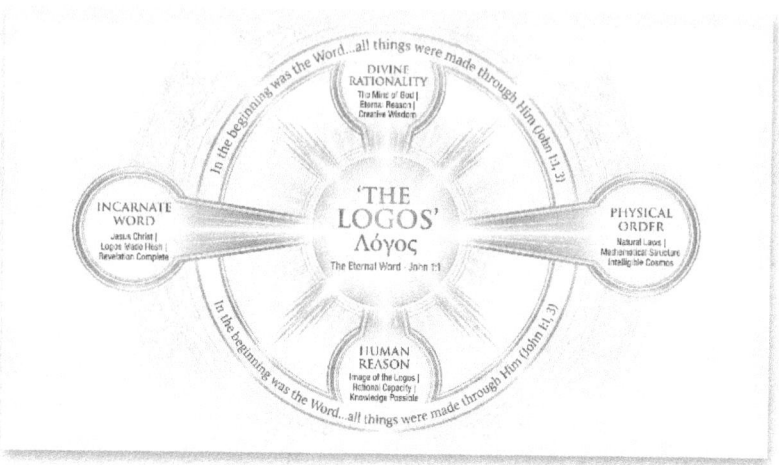

The Logos: The eternal Word through whom all things are made (John 1:1-3, Colossians 1:16-17). The Logos is the rational ground of physical order, the source of human reason, and revealed fully in Jesus Christ. (Chapters 10-14)

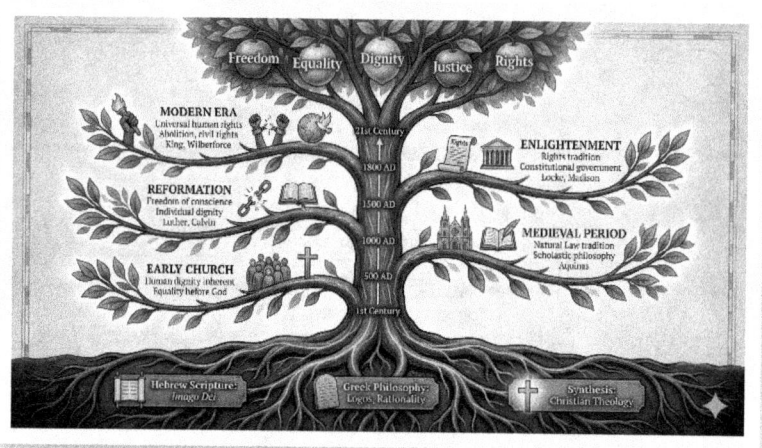

The Western Tradition: How Christian theological commitments to human dignity, equality, and freedom shaped the development of Western political and moral thought. (Chapters 19-21)

The Integrated Synthesis: How physics, metaphysics, and theology converge on a coherent vision of reality grounded in the eternal I AM. Each discipline illuminates the others without reduction or conflict. (Introduction, Conclusion, Part X)

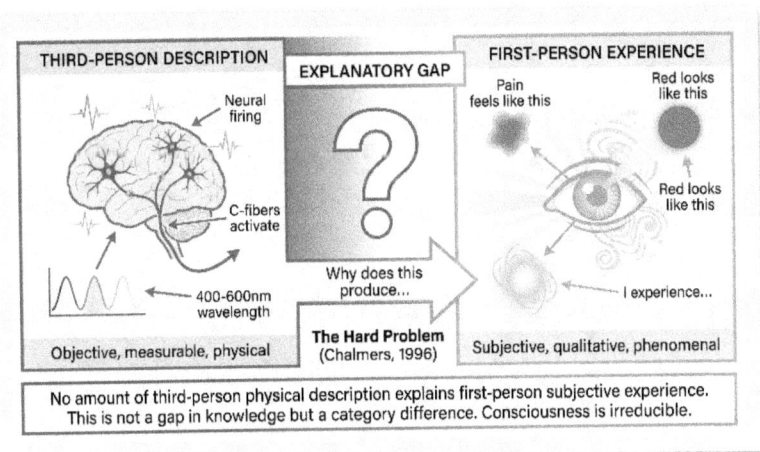

The Hard Problem of Consciousness: Why physical brain processes (third-person) cannot fully explain subjective experience (first-person). This points toward something beyond mechanism. (Chapter 7)

BIBLIOGRAPHY

This bibliography includes the major scientific, philosophical, theological, and historical works referenced throughout the book. It also provides a curated selection of foundational and recommended sources in cosmology, quantum theory, information science, classical theology, Logos studies, and Western political thought.

PRIMARY SOURCES - ANCIENT & CLASSICAL

Aquinas, Thomas. Summa Theologiae. Translated by the Fathers of the English Dominican Province. New York: Benziger Brothers, 1947.

Aquinas, Thomas. Summa Contra Gentiles.

Aristotle. De Anima (On the Soul).

Aristotle. Metaphysics. Translated by W. D. Ross. Oxford: Clarendon Press, 1924.

Athanasius. On the Incarnation. 4th century.

Augustine. Confessions. Translated by Henry Chadwick. Oxford: Oxford University Press, 1991.

Augustine. On the Spirit and the Letter. 412-415 AD.

Heraclitus. Fragments. Translated by T. M. Robinson. Toronto: University of Toronto Press, 1987.

John of Damascus. Exposition of the Orthodox Faith. Translated by S. D. F. Salmond. Edinburgh: T&T Clark, 1898.

Locke, John. Two Treatises of Government. Edited by Peter Laslett. Cambridge: Cambridge University Press, 1988.

Luther, Martin. The Freedom of a Christian. 1520.

Proclus. Elements of Theology. Translated by E. R. Dodds. Oxford: Clarendon Press, 1963.

Zeno of Citium. Fragments. In Lives of the Eminent Philosophers by Diogenes Laertius. Translated by Robert Drew Hicks. Cambridge: Harvard University Press, 1925.

PHYSICS & COSMOLOGY

Bekenstein, Jacob. "Black Holes and Entropy." Physical Review D 7 (1973): 2333-2346.

Bousso, Raphael. "The Holographic Principle." Reviews of Modern Physics 74 (2002): 825-874.

Einstein, Albert. Relativity: The Special and the General Theory. New York: Crown Publishers, 1961.

Ellis, George F. R., and Stephen Hawking. The Large Scale Structure of Space-Time. Cambridge: Cambridge University Press, 1973.

Greene, Brian. *The Elegant Universe: Superstrings, Hidden Dimensions, and the Quest for the Ultimate Theory*. New York: W. W. Norton, 1999.

Guth, Alan H. *The Inflationary Universe*. Reading, Massachusetts: Addison-Wesley, 1997.

Hawking, Stephen. *A Brief History of Time*. New York: Bantam Books, 1988.

Hawking, Stephen, and Leonard Mlodinow. *The Grand Design*. New York: Bantam Books, 2010.

Heisenberg, Werner. *Physics and Philosophy*. New York: Harper and Row, 1958.

Landauer, Rolf. "Information is Physical." *Physics Today* 44, no. 5 (1991): 23-29.

Maldacena, Juan. "The Large-N Limit of Superconformal Field Theories and Supergravity." *Advances in Theoretical and Mathematical Physics* 2 (1998): 231-252.

Maxwell, James Clerk. *A Treatise on Electricity and Magnetism*. Oxford: Clarendon Press, 1873.

Penrose, Roger. *The Road to Reality: A Complete Guide to the Laws of the Universe*. New York: Knopf, 2005.

Poincaré, Henri. *Science and Hypothesis*. New York: Dover Publications, 1952.

Prigogine, Ilya, and Isabelle Stengers. *Order Out of Chaos: Man's New Dialogue with Nature*. Bantam Books, 1984.

Ryu, Shinsei, and Tadashi Takayanagi. "Holographic Derivation of Entanglement Entropy." *Physical Review Letters* 96 (2006): 181602.

Schrödinger, Erwin. *What Is Life?* Cambridge: Cambridge University Press, 1944.

Shannon, Claude E. "A Mathematical Theory of Communication." *Bell System Technical Journal* 27, no. 3 (1948): 379-423.

Silk, Joseph. *The Big Bang*. 3rd ed. W. H. Freeman, 2001.

Tegmark, Max. *Our Mathematical Universe*. New York: Alfred A. Knopf, 2014.

Weinberg, Steven. *The First Three Minutes: A Modern View of the Origin of the Universe*. Basic Books, 1977.

Wheeler, John Archibald. "Information, Physics, Quantum: The Search for Links." *Proceedings of the Third International Symposium on Foundations of Quantum Mechanics* (1989): 354-368.

Wheeler, John Archibald. *At Home in the Universe*. New York: American Institute of Physics, 1994.

Wigner, Eugene. "The Unreasonable Effectiveness of Mathematics in the Natural Sciences." *Communications on Pure and Applied Mathematics* 13, no. 1 (1960): 1-14.

PHILOSOPHY & METAPHYSICS

Barrow, John D. *Pi in the Sky: Counting, Thinking, and Being*. Oxford University Press, 1992.

Chalmers, David J. - *The Conscious Mind* (1996)

Davies, Brian. *An Introduction to the Philosophy of Religion*. Oxford University Press, 2004.

Feser, Edward. *Aquinas: A Beginner's Guide*. Oneworld Publications, 2009.
Feser, Edward. *Scholastic Metaphysics: A Contemporary Introduction*. Editiones Scholasticae, 2014.
Kane, Robert, ed. *The Oxford Handbook of Free Will*. Oxford University Press, 2002.
Kuhn, Thomas S. *The Structure of Scientific Revolutions*. Chicago: University of Chicago Press, 1962.
MacIntyre, Alasdair. *After Virtue: A Study in Moral Theory*. University of Notre Dame Press, 1981.
Nagel, Thomas. *Mind and Cosmos: Why the Materialist Neo-Darwinian Conception of Nature Is Almost Certainly False*. Oxford University Press, 2012.
Nietzsche, Friedrich. *The Gay Science*. 1882/1974.
Parfit, Derek. *Reasons and Persons*. Oxford University Press, 1984.
Plantinga, Alvin. *God, Freedom, and Evil*. Eerdmans, 1974.
Plantinga, Alvin. *Where the Conflict Really Lies: Science, Religion, and Naturalism*. Oxford University Press, 2011.
Pruss, Alexander - *The Principle of Sufficient Reason* (2006)
Spaemann, Robert. *Persons: The Difference Between 'Someone' and 'Something'*. Translated by Oliver O'Donovan. Oxford University Press, 2006.
Steiner, Mark. *The Applicability of Mathematics as a Philosophical Problem*. Harvard University Press, 1998.
Stump, Eleonore. *Aquinas*. Routledge, 2003.
Swinburne, Richard. *Providence and the Problem of Evil*. Oxford University Press, 1998.
Swinburne, Richard. *Mind, Brain, and Free Will*. Oxford University Press, 2013.
Tononi, Giulio. "An Information Integration Theory of Consciousness." BMC Neuroscience 5, no. 42 (2004).
Van Inwagen, Peter. *Metaphysics*. Boulder: Westview Press, 1993.

THEOLOGY - SYSTEMATIC & HISTORICAL

Adams, Robert M. *Finite and Infinite Goods: A Framework for Ethics*. Oxford University Press, 1999.
Barth, Karl. *Church Dogmatics II/1: The Doctrine of God*. T&T Clark, 1956.
Barth, Karl. *Church Dogmatics II/2: The Doctrine of God*. T&T Clark, 1957.
Barth, Karl. *Church Dogmatics III/1: The Doctrine of Creation*. T&T Clark, 1958.
Barth, Karl. *Church Dogmatics IV/1: The Doctrine of Reconciliation*. T&T Clark, 1956.
Bonhoeffer, Dietrich. *Life Together*. Harper & Row, 1954.
Bonhoeffer, Dietrich. *The Cost of Discipleship*. Macmillan, 1959.
Brown, Raymond E. *The Gospel According to John (I-XII)*. Anchor Bible Commentary. Doubleday, 1966.
Clapp, Rodney. *A Peculiar People: The Church as Culture in a Post-Christian Society*. InterVarsity Press, 1996.

Copan, Paul, and William Lane Craig. Creation Out of Nothing: A Biblical, Philosophical, and Scientific Exploration. Baker Academic, 2004.

Craig, William Lane. Reasonable Faith: Christian Truth and Apologetics. Wheaton: Crossway, 2008.

Craig, William Lane, and James D. Sinclair. "The Kalam Cosmological Argument." In The Blackwell Companion to Natural Theology, edited by William Lane Craig and J. P. Moreland, 101-201. Oxford: Wiley-Blackwell, 2009.

Dolezal, James E. God Without Parts: Divine Simplicity and the Metaphysics of God's Absoluteness. Pickwick Publications, 2011.

Dunn, James D. G. Christology in the Making. 2nd ed. Eerdmans, 1989.

Finnis, John. Natural Law and Natural Rights. 2nd edition. Oxford University Press, 2011.

Foster, Richard J. Prayer: Finding the Heart's True Home. HarperSanFrancisco, 1992.

George, Robert P. In Defense of Natural Law. Oxford University Press, 1999.

Gunton, Colin E. The Triune Creator: A Historical and Systematic Study. Eerdmans, 1998.

Hauerwas, Stanley. Character and the Christian Life: A Study in Theological Ethics. Trinity University Press, 1975.

Hengel, Martin. Judaism and Hellenism. Fortress Press, 1974.

Hick, John. Evil and the God of Love. Macmillan, 1966.

Hoekema, Anthony A. Created in God's Image. Eerdmans, 1986.

Keener, Craig S. The Gospel of John: A Commentary. Baker Academic, 2003.

LaCugna, Catherine Mowry. God For Us: The Trinity and Christian Life. HarperSanFrancisco, 1991.

Lewis, C. S. The Abolition of Man. Oxford University Press, 1943.

Lewis, C. S. Mere Christianity. Macmillan, 1952.

May, Gerhard. Creatio Ex Nihilo: The Doctrine of 'Creation out of Nothing' in Early Christian Thought. T&T Clark, 1994.

McGuckin, John Anthony. The Westminster Handbook to Patristic Theology. Westminster John Knox Press, 2004.

Middleton, J. Richard. The Liberating Image: The Imago Dei in Genesis 1. Brazos Press, 2005.

Moltmann, Jürgen. The Crucified God: The Cross of Christ as the Foundation and Criticism of Christian Theology. SCM Press, 1974.

O'Donovan, Oliver. Resurrection and Moral Order. 2nd ed. Eerdmans, 1994.

Porter, Jean. Natural and Divine Law: Reclaiming the Tradition for Christian Ethics. Eerdmans, 1999.

Ratzinger, Joseph (Pope Benedict XVI). Introduction to Christianity. San Francisco: Ignatius Press, 2004.

Ratzinger, Joseph (Pope Benedict XVI). In the Beginning: A Catholic Understanding of the Story of Creation and the Fall. Grand Rapids: Eerdmans, 1995.

Stott, John R. W. The Cross of Christ. InterVarsity Press, 1986.

Thomas, Owen. The Thomistic Tradition. Notre Dame: University of Notre Dame Press, 1981.

Torrance, Thomas F. Divine and Contingent Order. Oxford University Press, 1981.
Torrance, Thomas F. The Mediation of Christ. Revised edition. Helmers & Howard, 1992.
Torrance, Thomas F. The Trinitarian Faith: The Evangelical Theology of the Ancient Catholic Church. T&T Clark, 1992.
Torrance, Thomas F. The Christian Doctrine of God, One Being Three Persons. T&T Clark, 1996.
Volf, Miroslav. Exclusion and Embrace: A Theological Exploration of Identity, Otherness, and Reconciliation. Abingdon Press, 1996.
Volf, Miroslav. After Our Likeness: The Church as the Image of the Trinity. Eerdmans, 1998.
Weinandy, Thomas G. Does God Suffer? University of Notre Dame Press, 2000.
Willard, Dallas. The Divine Conspiracy: Rediscovering Our Hidden Life in God. HarperSanFrancisco, 1998.
Wolterstorff, Nicholas. Divine Discourse: Philosophical Reflections on the Claim that God Speaks. Cambridge University Press, 1995.
Wolterstorff, Nicholas. Justice: Rights and Wrongs. Princeton University Press, 2008.
Wright, N. T. The Resurrection of the Son of God. Minneapolis: Fortress Press, 2003.
Wright, N. T. Simply Christian. New York: Harper One, 2006.
Wright, N. T. Surprised by Hope: Rethinking Heaven, the Resurrection, and the Mission of the Church. HarperOne, 2008.
Zizioulas, John D. Being as Communion: Studies in Personhood and the Church. St Vladimir's Seminary Press, 1985.

BIBLICAL STUDIES & ANCIENT LANGUAGES

Barr, James. The Semantics of Biblical Language. Oxford University Press, 1961.
Boman, Thorleif. Hebrew Thought Compared with Greek. SCM Press, 1960.
Kass, Leon R. The Beginning of Wisdom: Reading Genesis. Free Press, 2003.
Kirk, G. S., J. E. Raven, and M. Schofield. The Presocratic Philosophers. 2nd ed. Cambridge University Press, 1983.
Levenson, Jon D. Creation and the Persistence of Evil: The Jewish Drama of Divine Omnipotence. Princeton University Press, 1988.
Tsumura, David T. The Earth and the Waters in Genesis 1 and 2: A Linguistic Investigation. Sheffield Academic Press, 1989.
Walton, John H. Genesis 1 as Ancient Cosmology. Eisenbrauns, 2001.

SCIENCE & RELIGION

Carroll, Sean. The Big Picture: On the Origins of Life, Meaning, and the Universe Itself. New York: Dutton, 2016.
Davies, Paul. The Mind of God: The Scientific Basis for a Rational World. New York: Simon and Schuster, 1992.

Davies, Paul, and Niels Henrik Gregersen, eds. Information and the Nature of Reality: From Physics to Metaphysics. Cambridge University Press, 2010.

Dodds, Michael J. Unlocking Divine Action: Contemporary Science and Thomas Aquinas. Catholic University of America Press, 2012.

Jaki, Stanley L. The Road of Science and the Ways to God. Chicago: University of Chicago Press, 1978.

Polkinghorne, John. Belief in God in an Age of Science. Yale University Press, 1998.

Polkinghorne, John. Exploring Reality: The Intertwining of Science and Religion. Yale University Press, 2005.

HISTORY, POLITICAL PHILOSOPHY & WESTERN CIVILIZATION

Collins, Robin. "The Teleological Argument: An Exploration of the Fine-Tuning of the Universe." In The Blackwell Companion to Natural Theology, edited by William Lane Craig and J. P. Moreland, 202-281. Wiley-Blackwell, 2009.

Dreisbach, Daniel L., and Mark David Hall, eds. The Sacred Rights of Conscience: Selected Readings on Religious Liberty and Church-State Relations in the American Founding. Liberty Fund, 2014.

Inwood, Brad, and Lloyd P. Gerson, eds. The Stoics Reader: Selected Writings and Testimonia. Hackett, 2008.

Jonas, Hans. The Gnostic Religion: The Message of the Alien God and the Beginnings of Christianity. 2nd edition. Beacon Press, 1958.

King, Martin Luther, Jr. "Letter from Birmingham Jail." 1963.

Long, A. A. Hellenistic Philosophy: Stoics, Epicureans, Sceptics. 2nd ed. University of California Press, 1986.

Madison, James. "Federalist No. 51." In The Federalist Papers. 1788.

Sanchez-Sorondo, Marcelo. "The Importance of the Salamanca School." In Natural Law and Human Dignity, edited by Ernest Fortin. Washington, D.C.: University Press of America, 1990.

Siedentop, Larry. Inventing the Individual: The Origins of Western Liberalism. Belknap Press of Harvard University Press, 2014.

Stark, Rodney. The Victory of Reason: How Christianity Led to Freedom, Capitalism, and Western Success. Random House, 2005.

Witte, John, Jr. Religion and the American Constitutional Experiment. 2nd edition. Westview Press, 2000.

NOTES ON PART VIII Q&A SCRIPTURE REFERENCES

Part VIII (Questions 1-100) contains extensive Scripture references integrated within the text. These references are intentionally not included in this bibliography as they:

- Stand on Scripture's authority alone

BIBLIOGRAPHY

- Are formatted as inline citations within answers (e.g., References: Psalm 139:1-4; Hebrews 4:13)
- Serve to guide further biblical study rather than cite secondary sources

All Scripture references in Part VIII can be found in any standard Bible translation (ESV, NASB, NABRE, KJV, etc.)..

GLOSSARY

TECHNICAL DEFINITIONS ACROSS ALL DISCIPLINES

PHYSICS

The terms in this section come from modern theoretical and experimental physics. Their definitions are given with strict disciplinary accuracy to prevent confusion with metaphysical, theological, or informational meanings.

Physics describes quantitative structures, behaviors, and interactions within spacetime. It does not describe metaphysical causes, divine action, consciousness, or meaning.

This group defines physics in its own domain, ensuring conceptual purity.

Physics (as a Discipline)

The empirical and mathematical study of the physical universe, including matter, energy, spacetime, fields, and the fundamental interactions that govern them.

Physics seeks predictive and testable models expressed in mathematical form. It does not address metaphysical necessity, value, purpose, or the existence of God.

Physical Law

A mathematically expressed regularity describing how physical systems behave.

In physics:

- Laws do not dictate behavior
- Laws do not possess causal power
- Laws do not explain metaphysical necessity

They describe consistent patterns in nature.

System

A defined portion of the physical world under analysis.

A "system" in physics is:

- context-dependent
- observer-defined
- not an ontological entity

This term differs from metaphysical or theological notions of "system."

State

A complete specification of the variables describing a system at a given moment. A "state" is mathematical, not metaphysical.

Observable

A measurable quantity associated with a physical system (e.g., position, momentum, spin).

Not to be confused with metaphysical or semantic "observation."

Observer (Physics)

Anything; a device, detector, or environment that interacts with a system such that information becomes definite.

In physics, an observer:

- does not need to be conscious
- is not a metaphysical or theological subject
- is any physical system capable of interaction

This prevents the mistake of equating quantum observation with human consciousness.

Interaction

A physical process by which systems exchange energy, momentum, or information.

In quantum mechanics, interaction often leads to entanglement or state update.

Measurement

A physical interaction that produces a definite outcome for an observable.

Measurement:

- is not identical to perception
- does not imply consciousness
- is a physical event
- typically involves decoherence

Cause (Physics)

In physics, "cause" refers to:

- dynamical in evolution
- initial conditions
- forces or interactions

Physics never uses "cause" to refer to metaphysical origins or divine action.

Time (Physical)

A dimension in which events occur in sequence.

In physics, time:

- is modeled mathematically
- is measured by clocks
- does not imply metaphysical becoming
- may differ across reference frames (relativity)

This differs sharply from metaphysical time or theological eternity.

Energy

A conserved physical quantity representing the capacity to perform work or cause motion.

Energy:

- is not a metaphysical principle
- is not a mystical force
- is not identical to life, consciousness, or spirit

Field

A physical entity assigning values (scalars, vectors, tensors) to points in spacetime. Modern physics views fields, not particles, as fundamental.

This term must not be confused with:

- metaphysical "fields of being"
- theological "spiritual fields"
- information-theoretic abstractions

Particle

A localized excitation of an underlying field (quantum field). Particles:

- are not tiny billiard balls
- are not metaphysical atoms
- do not possess fixed trajectories in quantum theory

This definition prevents classical misunderstandings.

Quantum System

A system described by quantum mechanics, characterized by:

- superposition
- probabilistic outcomes

- state vectors
- operators
- Hilbert space representation

Quantum systems do not behave according to classical intuition.

Superposition
A quantum state in which multiple possible outcomes coexist mathematically until measurement.

Superposition:

- is not a physical mixture
- is not "both at once" in a classical sense
- is not metaphysical indeterminacy
- is a mathematical representation of probability amplitudes

Probability Amplitude
A complex number encoding the likelihood of outcomes. Amplitude ≠ probability. Probability is the square of amplitude magnitude.

Quantum Indeterminacy
Outcomes cannot be predicted with certainty; only probabilities can be known. This is epistemic and structural — not metaphysical randomness.

QUANTUM MECHANICS CORE CONCEPTS

Wave Function (Ψ)
A mathematical function encoding all possible states of a quantum system.

The wave function:

- exists in Hilbert space, not physical space
- is a tool for calculating probabilities

- is not a physical wave
- is not consciousness
- is not metaphysical potentiality

Hilbert Space

An abstract mathematical space in which quantum states are represented as vectors.

Important distinctions:

- Not a physical place or medium
- Not spacetime
- Not a metaphysical realm
- Purely mathematical

State Vector

A vector in Hilbert space representing the quantum system's full possible configuration.

State vectors encode:

- amplitudes
- probabilities
- observable structure

They do not encode meaning, consciousness, or metaphysical essence.

Operator

A mathematical object acting on state vectors to produce measurable predictions.

Different operators correspond to different observables (position, momentum, spin).

Operators:

- are not "forces"
- do not act in spacetime
- exist purely in the mathematical formalism

Eigenvalue

A possible outcome of measuring an observable.
Each eigenvalue corresponds to:

- an eigenstate
- a specific measurement result

Eigenvalues are not metaphysical essences or values — they are numerical outcomes.

Eigenstate

A state associated with a definite value of an observable.
When a system is in an eigenstate of an operator:

- measurement yields a certain result
- no probability distribution remains

This helps avoid confusion about determinacy and probability.

Measurement Problem

The unresolved question of how and why quantum possibilities become definite outcomes.

Key clarifications:
The measurement problem is philosophical, not simply technical

- It does not require consciousness
- It does not imply metaphysical idealism

Different interpretations propose different mechanisms. This book does not endorse any nonphysical interpretation.

Wave Function Collapse (Interpretive Term)

"Collapse" refers to the transition from a superposed description to a definite one after measurement.

Important notes:

- Collapse is not a physical explosion

- Collapse is not caused by consciousness in mainstream physics
- Collapse is not guaranteed to represent a physical event
- Collapse is often treated as an information update, not a metaphysical act

This book uses collapse informationally, not metaphysically.

Decoherence

The process by which quantum superposition effectively disappears due to environmental interaction.

Decoherence:

- explains classical behavior
- does not solve the measurement problem
- does not require mind or observer consciousness
- is entirely physical

It prevents misinterpretation that "observation = mind power."

Entanglement

A correlation between systems such that the state of one cannot be described independently of the other.

Clarifications:

- Entanglement is not telepathy
- Entanglement is not faster-than-light communication
- Entanglement is not evidence for mystical unity
- Entanglement is not metaphysical unity

Entanglement is a mathematical relation in quantum mechanics

Nonlocal Correlation

The statistical relationship between entangled particles.

This does not violate:

- relativity

- causality
- light-speed constraints

Only information about correlations appears nonlocal — not usable communication.

Quantum Field

A field that permeates spacetime; particles are excitations of these fields.

Clarifications:

- Fields ≠ metaphysical fields
- Fields ≠ divine presence
- Fields ≠ consciousness
- Fields ≠ informational abstractions

Quantum fields are mathematical structures with physical manifestations.

Vacuum State

The lowest-energy state of a quantum field, containing fluctuations due to the uncertainty principle.

- Vacuum ≠ nothingness.
- Vacuum ≠ metaphysical non-being.

Zero-Point Energy

The lowest possible energy of a quantum system. Not a "spiritual force," not evidence of consciousness, not metaphysical being.

Uncertainty Principle (Heisenberg)

Certain pairs of observables (e.g., position and momentum) cannot both be known with unlimited precision.

Key clarification:

- This is not due to measurement disturbance
- It is a structural feature of quantum systems
- It does not imply metaphysical randomness

- It does not imply subjectivity

Quantum Potentiality vs. Metaphysical Potency

- Quantum potentiality = mathematical representation of probabilities
- Metaphysical potency = ontological capacity for becoming

They are not the same, and this text clearly distinguishes them.

MATHEMATICAL STRUCTURES, SYMMETRIES, AND RELATIVITY

Path Integral (Feynman Integral)

A formulation of quantum mechanics in which a particle's behavior is calculated by summing over all possible paths it could take, each weighted by a complex phase factor.

Clarifications:

- Not every path is literally taken
- It is a computational framework, not a physical ontology
- Does not imply infinite realities or "many worlds" unless interpreted that way

Feynman Diagram

A pictorial tool for calculating interactions in quantum field theory. It represents terms in a perturbation series, not literal particle trajectories.

Misunderstandings prevented:

- These diagrams are not pictures of reality
- Lines do not represent actual paths
- Virtual particles are not real particles

Virtual Particle

A term from quantum field calculations indicating internal mathematical elements of interaction diagrams.

Virtual particles:

- do not exist as real entities
- cannot be observed
- have no physical trajectory
- are computational artifacts

Symmetry (Physics)

A transformation under which the laws of physics remain unchanged.

Examples:

- rotational symmetry
- translational symmetry
- gauge symmetry
- Lorentz symmetry

Symmetry ≠ metaphysical perfection or beauty unless used analogically.

Noether's Theorem

A foundational result connecting symmetries with conservation laws.

Examples:

- time symmetry → energy conservation
- spatial symmetry → momentum conservation
- gauge symmetry → charge conservation

This is a mathematical necessity, not a metaphysical one.

Gauge Theory

A type of field theory in which symmetries determine allowed interactions.

Examples:

- electromagnetism (U(1) gauge symmetry)
- weak force (SU(2))
- strong force (SU(3))

Gauge fields are mathematical structures dictating interaction behavior.

Gauge Boson

Particles mediating forces in gauge theories.
For example:

- photon
- W and Z bosons
- gluons

They are field excitations, not metaphysical "messengers."

Locality (Physics)

Physical influences propagate no faster than light; interactions occur through fields and forces that respect spacetime structure.
Important:

- Quantum correlations (entanglement) do not violate locality in a causal sense.

Lorentz Invariance

One of the cornerstones of relativity: the laws of physics are the same for all observers moving at constant velocity relative to one another. Not equivalent to metaphysical claims about perspective or consciousness.

Spacetime (Relativistic)

A four-dimensional manifold containing:

- three spatial dimensions
- one temporal dimension

This is a physical model, not a metaphysical description of being.

Worldline
The trajectory of an object through spacetime.
Not a literal path the object "experiences"; it is a geometric description.

Proper Time
The time measured by a clock moving along a worldline. Different observers may measure different times depending on motion and gravity.

Important:

- This has no connection to metaphysical or subjective time.

Curvature (General Relativity)
Gravity is modeled as curvature of spacetime caused by mass-energy.

Key clarifications:

- Curvature is not a force
- Curvature is geometric, not metaphysical
- Curvature does not imply intention or design

Einstein Field Equations (EFE)
The central equations describing how mass-energy determines spacetime curvature. Not metaphysical necessity — purely physical description.

Metric Tensor
The mathematical structure defining distances and intervals in spacetime.
This is not a "field of meaning" or metaphysical order.

Geodesic
The path of least action (or "straightest path") through spacetime.

Objects follow geodesics unless acted on by non-gravitational forces.

Cosmic Speed Limit (c)

The maximum speed at which information or causal influence can propagate.

Clarifications:

- Quantum entanglement does not violate this
- Spiritual or metaphysical realities are not subject to this law

Rest Mass vs. Relativistic Mass:

- Rest mass = invariant property of a particle
- Relativistic mass (older term) increases with speed; not used in modern physics.

Important for avoiding outdated explanations.

Stress-Energy Tensor

Describes distribution of energy, momentum, pressure, and density in spacetime.

Important distinction:

- This is physical structure, not metaphysical "energy."

FOUNDATIONS OF COSMOLOGY & LARGE-SCALE PHYSICAL STRUCTURE: (NOTE: THESE TERMS BELONG TO PHYSICS, NOT METAPHYSICS OR THEOLOGY.)

Big Bang Model

The standard cosmological model in which the universe expands from a hot, dense initial state approximately 13.8 billion years ago.

Clarifications:

- The Big Bang is not an explosion in space
- It is an expansion of space
- It does not describe the ultimate origin of existence
- It does not explain why the universe exists (metaphysical question)

Singularity (Mathematical)

A point where a physical model breaks down (infinite density, infinite curvature).

This is not:

- a literal point in reality
- a metaphysical nothingness
- a moment of creation

It marks the limit of current physical theories.

Inflation (Cosmic Inflation)

A period of extremely rapid exponential expansion in the early universe, proposed to explain:

- the flatness problem
- the horizon problem
- the absence of magnetic monopoles
- the uniformity of the cosmic microwave background (CMB)

Inflation is a physical hypothesis, not a metaphysical claim about purpose or design.

Cosmic Microwave Background (CMB)

The faint thermal radiation left over from the recombination era (~380,000 years after the Big Bang).

It serves as:

- a snapshot of early cosmic conditions
- evidence for the Big Bang model

- a probe for fluctuations and structure formation

But it does not tell us why the universe exists.

Horizon (Cosmological)

A boundary beyond which events cannot affect an observer because light hasn't had time to reach them.

Types include:

- particle horizon
- event horizon
- Hubble sphere

These horizons are physical, not metaphysical.

Initial Conditions

The very specific starting parameters required for the universe to evolve as it does:

- extremely low entropy
- uniform temperature
- tiny quantum fluctuations

Physics describes what these conditions were, not why they were that way.

Low-Entropy Past (Past Hypothesis)

The universe began in a state of extraordinarily low entropy, giving rise to:

- the arrow of time
- increasing complexity
- structure formation

This asymmetry is empirical, not metaphysical necessity.

Structure Formation

The development of:

- galaxies
- stars
- planets
- cosmic filaments

Arising from the amplification of tiny density fluctuations in the early universe. No metaphysical teleology is implied.

Dark Matter

A form of matter that does not interact with light but exerts gravitational influence.

Dark matter:

- is not mystical
- is not spiritual
- is not metaphysical
- is a physical component inferred from observations

Dark Energy

Dark Energy: The component driving accelerated expansion, modeled as a vacuum energy density (Λ) in standard cosmology. The physical mechanism remains an active area of research.

It is:

- not consciousness
- not divine energy
- not informational energy

It is a parameter in Einstein's equations and a subject of ongoing research.

Cosmological Constant (Λ)

A term in Einstein's field equations representing the energy density of empty space (vacuum energy).

This is mathematical, not metaphysical.

Baryon Acoustic Oscillations

Periodic fluctuations in density within the early universe's plasma.

These are:

- physical phenomena
- used for measuring cosmic distances
- not metaphysical rhythms or patterns

Nucleosynthesis (Big Bang Nucleosynthesis)
The formation of the first atomic nuclei (hydrogen, helium, trace lithium) within minutes of the Big Bang. Not a metaphysical or mystical act of creation.

Fine Structure Constant (α)
A dimensionless constant measuring the strength of electromagnetic interaction.

It is:

- not adjustable by metaphysical forces
- not a sign of design unless interpreted theologically
- a physical parameter used in calculations

Hubble Expansion
The observation that galaxies recede from each other at speeds proportional to their distance.

Expansion:

- does not imply a central point
- does not imply metaphysical expansion
- is not an explosion

It is a geometric property of spacetime.

Cosmic Scale Factor (a(t))
A function describing how distances between two points in the universe change over time. This is mathematical, not metaphysical.

ΛCDM Model

The current standard model of cosmology, including:

- Λ (cosmological constant)
- C (cold)
- D (dark)
- M (matter)

It is descriptive, not ultimate.

COSMOLOGY
STRUCTURE, EXPANSION, AND FOUNDATIONS

Cosmology (Physical Cosmology)
The scientific study of the large-scale structure, dynamics, and evolution of the universe.
Cosmology does not:

- investigate metaphysical origins
- answer why the universe exists
- identify divine causality
- study personal meaning

It describes the universe's behavior, not its ultimate grounding.
Cosmic Expansion
The increase of distances between galaxies over time due to the stretching of spacetime itself.
Clarifications:

- not an explosion
- no center of expansion
- applies to large scales, not bound systems (solar systems, galaxies)

Scale Factor (a(t))

A mathematical function describing how spatial distances change over cosmic time. It is not a physical substance, force, or metaphysical process.

Friedmann–Lemaître–Robertson–Walker (FLRW) Metric

A cosmological solution to Einstein's field equations assuming:

- homogeneity
- isotropy

It models the universe on large scales.

Spatial Curvature

The geometric curvature of space at cosmic scales, which can be:

- positive (closed)
- zero (flat)
- negative (open)

This curvature is physical, not metaphysical.

Hubble Parameter (H(t))

The rate of cosmic expansion at a given time.
Not a force, not an "expansion field," just a mathematical derivative.

Redshift (z)

The stretching of light toward longer wavelengths as the universe expands. Redshift is empirical evidence of expansion, not metaphysical movement.

Comoving Distance

A distance measure corrected for cosmic expansion, holding spatial coordinates fixed relative to the expanding grid. Useful for comparing large-scale structures.

Lookback Time

The time difference between now and the moment when the observed light was emitted.

Cosmic observation is always observation of the past.

Cosmic Time
A universal time coordinate used in FLRW cosmology. It is:

- not metaphysical time
- not psychological time
- not divine eternity

Cosmological Principle
The assumption that the universe is homogeneous and isotropic on large scales.
A methodological principle, not a metaphysical claim.

Homogeneity
Uniformity of matter distribution on sufficiently large scales.

Isotropy
Uniformity when viewed in every direction.

HORIZONS, LIMITS, AND LARGE-SCALE STRUCTURE

Particle Horizon
The maximum distance from which light could have traveled to an observer since the Big Bang.
Important:

- This is a physical boundary of observability, not existence.

Event Horizon (Cosmological)
A boundary beyond which events will never be observable, even in the far future. Different from a black hole event horizon, though conceptually similar.

Hubble Radius

The distance at which recession velocity equals the speed of light. Not a physical barrier.

Cosmological Singularity (Model Breakdown)

A location where equations diverge or fail. This does not mean the universe was a point, metaphysical nothing, or a literal instant of creation. It is where current physics becomes incomplete. The theological interpretation sees this as the boundary where divine creation begins, but this is metaphysical interpretation, not physical explanation.

This does not mean:

- the universe was a point
- metaphysical nothing
- a literal instant of creation

It is where current physics becomes incomplete.

Cosmic Inflation (Revisited)

A period of rapid exponential expansion driven by an inflation field or effective potential.

Important:

- Inflation is a physical hypothesis, not metaphysics.

Reheating

The process by which energy from the inflation field converts into particles and radiation as inflation ends. Not a creation event — a physical transition.

Quantum Fluctuations (Cosmological)

Small variations in density during inflation that seed large-scale structure. Not metaphysical randomness.

Cosmic Web

The large-scale structure of the universe consisting of:

- filaments
- clusters

- voids

Arising from gravitational amplification of density fluctuations.

Baryonic Matter
Ordinary matter composed of protons, neutrons, and electrons. Not metaphysically privileged.

Non-Baryonic Matter
Matter not made of standard atomic particles (e.g., dark matter). This is physical, not spiritual or metaphysical.

Cosmic Variance
The statistical uncertainty arising from having only one observable universe. A limitation of measurement, not metaphysical uniqueness.

MULTIVERSE HYPOTHESES, HOLOGRAPHY, AND CONCEPTUAL BOUNDARIES: MULTIVERSE (PHYSICAL HYPOTHESES)

Various models proposing multiple universes arising from:

- inflationary bubbles
- quantum branching
- string theory landscapes

Clarifications:

- Multiverse models are physical hypotheses
- They do not solve metaphysical questions
- They do not eliminate the need for explanation.
- They are not equivalent to infinite realities in philosophy

Anthropic Principle (Weak / Strong)
Weak anthropic principle: Observations are biased by the fact that only universes compatible with observers can be observed.

Strong anthropic principle: More speculative; used carefully. Anthropic reasoning is methodological, not metaphysical or theological.

Holographic Principle (Cosmology Context)

The proposal that the information content of a region of space is proportional to its surface area, not its volume.

Important:

- This is mathematical, not ontological
- It does not imply the universe is an illusion
- It does not imply simulation theory
- It does not equate information with intelligence

Analogies involving information or holography are not claims that the universe is a computer simulation or illusion.

Black Hole Thermodynamics (Cosmology Connection)

The laws relating:

- entropy
- area
- surface gravity
- temperature

Important for understanding holography, but not metaphysical meaning.

Apparent Horizon / Event Horizon (Black Hole)

Boundaries defined by light paths and gravitational geometry. Not metaphysical or symbolic boundaries.

Observer-Dependence (Relativity)

Certain quantities depend on the observer's motion or position, but:

- reality is not subjective
- truth does not depend on belief
- consciousness does not shape spacetime

This prevents misreading relativity as philosophical relativism.

Cosmic Fine-Tuning (Physics Interpretation)

The observation that physical parameters fall within narrow ranges allowing structure to form.

Fine-tuning is:

- empirical
- descriptive
- not automatically theological
- not metaphysically necessary

Interpretation depends on philosophical framework.

Holographic Cosmology (Caution)

Applications of holography to cosmological models remain speculative and mathematical.

This book:

- uses holography for structural analogy
- does not treat holography as metaphysical truth
- avoids overextension of the principle

INFORMATION THEORY
CORE DEFINITIONS & FOUNDATIONAL CONCEPTS

Information (General Definition)
A measure of distinctions—the differentiation between possible states of a system.

Information, in the scientific sense, is:

- structural
- formal
- measurable
- relational

Information does not imply:

- meaning
- intentionality
- intelligence
- consciousness
- purpose

This distinction is crucial. Meaning, intentionality, and consciousness belong to subjects, not to physical information itself.

Shannon Information (Shannon Entropy)

A quantitative measure of uncertainty or "surprise" in a set of possible messages.

Important:

- It is mathematical, not semantic
- It measures unpredictability, not meaning
- It does not refer to consciousness
- It is distinct from thermodynamic entropy, though related formally

Bit (Binary Digit)

The fundamental unit of classical information; represents one of two states (0 or 1).

Clarifications:

- Not a metaphysical atom
- Not a physical particle (though it can be encoded physically)
- Not inherently meaningful

Channel

A medium or pathway through which information is transmitted.

Examples:

- wire
- fiber optic line
- electromagnetic signal

A channel is a physical or logical construct, not a metaphysical concept.

Noise

Anything that interferes with the accurate transmission of information.

Noise is:

- physical
- statistical
- not metaphysical chaos
- not moral disorder

Signal

The portion of transmitted information intended to be conveyed, distinguished from noise.

A signal does not necessarily carry meaning; it carries structured distinctions.

Message

An ordered set of symbols or states transmitted through a channel.

Important:

- "Message" in information theory does not imply intention or consciousness.

Code

A system of rules for representing information.
Examples include:

- ASCII
- Morse code
- Huffman coding

Codes are formal, not metaphysical or biological unless specified.

COMPLEX INFORMATION, ALGORITHMS, AND PHYSICAL INFORMATION

Algorithm

A step-by-step procedure or rule set for solving a problem or transforming information.

An algorithm:

- is formal
- is not conscious
- is not teleological
- has no intrinsic meaning

Algorithmic Complexity (Kolmogorov Complexity)

A measure of how compressible information is; the length of the shortest algorithm that reproduces a dataset.

- High complexity = low compressibility.
- Low complexity = high compressibility.

Not directly related to metaphysical or philosophical complexity.

Landauer's Principle

Erasing information requires energy.
Specifically:

- one bit erased $\rightarrow$ minimum energy cost of $kT \ln 2$

This connects thermodynamics and information but does not imply:

- information is physical substance
- information is metaphysical reality
- mind or intention plays a role

It's a physical consequence of computation.

Mutual Information

A measure of the statistical dependence between two sets of data or variables. It does not imply relationality in the metaphysical or personal sense.

Information Capacity

The maximum amount of information a system or channel can store or transmit.

Examples:

- black hole information capacity (~area/4 in Planck units)
- Shannon channel capacity

Not to be confused with meaning or knowledge.

Information Content

The degree to which a system distinguishes between possible states.

Information content ≠ semantic content.

Semantic Information (Outside Information Theory)

Meaningful information — used in linguistics or cognition.

Important:

- Semantic information is not part of Shannon's theory.

This book distinguishes semantic meaning from physical information.

PHYSICAL INFORMATION & QUANTUM INFORMATION

Physical Information

Information encoded in the physical state of a system.

Examples:

- spin

- momentum
- polarization
- position

Physical information is measurable and operational, not metaphysical.

Quantum Information
Information encoded in quantum systems, represented by:

- qubits
- state vectors
- density matrices

Quantum information has properties classical information does not, such as:

- superposition
- entanglement
- quantum no-cloning

Qubit (Quantum Bit)
The basic unit of quantum information.
A qubit:

- exists in a superposition of 0 and 1
- collapses into one state upon measurement
- is described by a vector in Hilbert space

A qubit is not a metaphysical or conscious entity.

Quantum Entropy (von Neumann Entropy)
Measures the uncertainty of a quantum state.
Related to classical Shannon entropy but distinct in formulation.

Quantum Mutual Information
Measures correlations between quantum systems. Does not

imply metaphysical interconnectedness.

No-Cloning Theorem
States that unknown quantum states cannot be perfectly copied. A purely mathematical consequence of linearity — not a metaphysical limit.

Quantum Channel
A medium through which quantum information is transmitted. It can:

- preserve entanglement
- suffer from decoherence
- experience noise

It is not a metaphysical conduit.

Holevo Bounds
The maximum amount of classical information extractable from a quantum system. Prevents over-interpretation of quantum states as "infinite information carriers."

INFORMATION IN PHYSICS & COSMOLOGY

Black Hole Information Problem
The question of whether information that falls into a black hole is lost or preserved.

This is:

- a physical puzzle
- not a metaphysical claim
- not evidence for God or meaning
- not resolved fully by current physics

Holographic Information
The idea that the information content of a region of space scales with its boundary area.

Important:

This is physical information, not semantic or metaphysical information.

Entropy–Information Relationship

In physics:

- Increasing entropy = increasing uncertainty
- Increasing information = decreasing uncertainty about system state

This book treats this relationship as physical, not metaphysical.

Information-Theoretic Universe (Caution)

The view that information is fundamental to physical reality.

Key Clarification:

The (IAM) does NOT claim:

- information is divine
- information is conscious
- information is metaphysical essence
- information is identical to Logos

Instead:

- information describes structure
- information participates in relationality
- metaphysics interprets why structure exists

METAPHYSICS

These terms come from the classical metaphysical tradition, primarily Aristotelian–Thomistic philosophy. They clarify the conceptual framework underlying this book's treatment of being, causality, order, and intelligibility. Each term is given in its rigorous academic meaning to prevent confusion with physics, information theory, or theological language.

Act (Actuality)
The fullness or realization of a thing's being. "Act" refers to what something is in its completed, actual state. It stands in contrast to potency.

Potency (Potentiality)
The capacity or possibility for becoming, changing, or taking on new states. Potency is not sameness with quantum "potential;" it is metaphysical, not physical. The "potential" in quantum mechanics concerns probability amplitudes, whereas metaphysical potency refers to ontological capacity; the two should not be conflated.

Act and Potency (Act–Potency Composition)
Every changeable being is composed of act (what it is already) and potency (what it can become).

This idea prevents contradictions and explains change without metaphysical incoherence.

Essence

"What a thing is" — its definable nature.

Essence is not physical structure; it is the intrinsic identity of a being.

Existence (Act of Existing)

"That a thing is" — the act by which essence is instantiated. Essence answers what something is; existence answers that it is.

Act of Being (Actus Essendi)

The deepest metaphysical act: the actuality of existence itself. Only God possesses this identically with His essence; created beings receive it. "Act of being' refers to the fundamental metaphysical actuality by which something exists at all, not a physical action.

Form

The organizing principle that makes a thing the kind of thing it is. Not a shape or structure in the physical sense, but the metaphysical "form of intelligibility."

Matter (Hyle)

The principle of potentiality in physical beings, that which can receive form. Not equivalent to "material particles;" this is metaphysical matter.

Substance

A being that exists in itself and not in another. Humans, animals, and atoms are substances.

Qualities and relations are not substances; they exist in substances.

Accident

A non-essential property inhering in a substance. For example: color, size, location, posture.

Accidents can change without altering essence.

Being (Ens)

Anything that exists in any way.

Being is not limited to physical existence; it includes mental, abstract, and divine being.

Contingent Being
A being that might not exist; its existence is received and dependent. All created beings fall under this category.

Necessary Being
A being that cannot not exist; its essence is existence. In classical metaphysics, only God fits this definition.

Subsistence (Subsistent Being)
A being that exists in a complete and independent manner. Human souls subsist; abstract objects do not.

Participation
The metaphysical relation by which finite beings receive their existence from the ultimate source of being.

Causality (Four Causes)
Metaphysics recognizes multiple layers of causation:

- Material cause — what something is made of
- Formal cause — what it is
- Efficient cause — what brings it about
- Final cause — its purpose or direction

Only efficient causes overlap with physical causation. The other causes are not physical.

Finality (Teleology)
Orientation toward an end, function, or outcome. This does not imply consciousness or intention unless explicitly stated.

Analogy of Being
The principle that words applied to God and creatures are neither purely literal nor purely metaphorical, but analogical. This prevents anthropomorphic errors.

Simplicity (Divine Simplicity)
God is not composed of parts, potentialities, or properties. This is a metaphysical doctrine, not a physical one.

Pure Act (Actus Purus)
A being with no potency — entirely actual.

Only God meets this description.

Composite Being

A being composed of act and potency, essence and existence, or form and matter. All created beings are composites.

Accidental vs. Essential Properties

- Essential: belong to what something is
- Accidental: can change without altering essence

This distinction clarifies nature vs. circumstance.

Intrinsic vs. Extrinsic Causality

- Intrinsic: arising from a being's nature
- Extrinsic: imposed from outside

Important in metaphysical and theological arguments.

Immaterial Reality

A reality not composed of physical matter but still real:

- intellect
- will
- rationality
- mathematical objects
- metaphysical principles
- God

Ontological Hierarchy

The distinction between levels of being:

- physical
- conceptual
- metaphysical
- divine

None of these are reducible to the others.

Potential Infinity vs. Actual Infinity
Created reality contains potential infinities (unending processes), but only God is actually infinite in being.
Real Distinction (Essence–Existence Distinction)
In creatures, essence and existence are distinct.
In God, they are identical.
Per Se vs. Per Accidens Causation

- Per se: ordered here-and-now (e.g., hand moves stick moves stone)
- Per accidens: temporal chains (e.g., father begets son)

This distinction is essential in classical arguments for God.
Efficient Cause (Metaphysical)
Not identical to a physical mechanism.
A metaphysical cause is deeper: it is the reason something exists or continues to exist.

CONSCIOUSNESS & PHILOSOPHY OF MIND: FOUNDATIONS OF CONSCIOUSNESS

Consciousness (General Definition)
The subjective, first-person awareness of experience.
Consciousness includes:

- qualia (felt experience)
- intentionality (aboutness)
- self-awareness
- unity of experience

This definition is philosophical and phenomenological, not physical or computational.
Consciousness is not:

- an information state

- a quantum process
- an emergent illusion
- a material property
- identical to neural activity

Neural activity correlates with consciousness but does not constitute it.

Subjectivity

The interior aspect of experience: the "what it is like" component of mental life.

Subjectivity is:

- private
- first-person
- irreducible to third-person description

This term is essential for distinguishing conscious awareness from physical or informational processes.

Qualia

The raw, subjective qualities of experience (e.g., redness, pain, taste).

Important distinctions:

- Qualia cannot be captured by descriptions of brain states
- Qualia are not computational states
- Qualia cannot be measured empirically
- Qualia are not reducible to information encoding

Intentionality

The capacity of mental states to be "about" something.
Intentionality:

- is not spatial direction
- is not intention in the sense of plans or goals

- is the directedness of thought toward objects, ideas, or states of affairs

This is a classical philosophical term originating in Scholastic and phenomenological traditions.

Self-Awareness (Reflexive Consciousness)
Awareness of oneself as a subject of experience.
Self-awareness:

- is not identical to memory
- is not self-modeling computation
- is not mere behavioral self-reference
- possesses an inward dimension not captured by physical description

Unity of Consciousness
The integration of diverse sensory, cognitive, and emotional processes into a single coherent field of awareness.

Unity is not:

- a computational integration
- a neural synchronization
- a mere functional convergence

It is a phenomenological unity.

THE NATURE OF MIND AND MENTAL STATES

Mind (Philosophical Definition)
The set of capacities enabling:

- thought
- reasoning
- awareness
- reflection

- intentionality

The mind is not reducible to:

- brain
- computation
- neural software
- physical information states

The distinction between mind and brain is foundational in philosophy.

Mental State

A state characterized by awareness, thought, or qualitative experience.

Mental states are:

- internally accessible
- phenomenologically defined
- not purely functional or computational

Cognitive State

A mental state involving thought, reasoning, memory, or judgment.

Cognition is:

- distinct from consciousness
- not identical to awareness
- not purely computational logic

Intentional State

A mental state involving reference to an object, idea, or condition. Examples:

- believing

- imagining
- desiring
- fearing
- perceiving

Intentional states are defined by their "aboutness," not physical location.

Propositional Attitude
A mental relation to a proposition (e.g., believing that, desiring that, hoping that). Not reducible to syntactic data or neural encodings.

THEORIES OF MIND (CLARIFICATIONS & BOUNDARIES)

Dualism (General Category)
The position that mind and matter are distinct in some fundamental way.

Forms include:

- substance dualism
- property dualism
- hylomorphic dualism

Dualism does not necessarily imply disembodied existence or anti-scientific thinking.

Physicalism
The view that everything that exists is ultimately physical.

- Physicalism does not automatically explain:
- consciousness
- qualia
- intentionality
- free will

Most variations struggle with the "hard problem" of consciousness.

Functionalism

The theory that mental states are defined by their functional roles, not their material composition.

Functionalism cannot account for:

- qualia
  - subjective experience
  - intrinsic intentionality

Emergentism

The idea that consciousness arises from complex physical systems.

Two major forms:

- Weak emergence: purely structural or functional
  - Strong emergence: new properties with causal powers

Emergence does not solve the explanatory gap.

Reductionism

The attempt to explain mental phenomena entirely in terms of physical processes. Reductionism is widely criticized for failing to explain subjective experience.

Non-Reductive Physicalism

Claims that mental states depend on but are not reducible to the physical.

Challenges:

- causal closure of physics
- nature of mental causation
- persistent hard problem

Panpsychism (Clarification)

The view that all matter has consciousness-like properties. This work does not adopt or imply panpsychism.

Panpsychism conflates:

- information
- structure
- relationality

with conscious awareness. This manual clarifies strict boundaries to prevent that confusion.

CONSCIOUSNESS, INFORMATION, AND MISINTERPRETATION PREVENTION

Consciousness ≠ Information

Information describes distinctions within a system. Consciousness is experience, which cannot be reduced to distinctions or data.

Consciousness ≠ Computation

Computations manipulate symbols syntactically. Consciousness is semantic and experiential.

Consciousness ≠ Quantum Mechanics

Quantum processes:

- have no subjectivity
- have no qualitative character
- have no intrinsic awareness

Any analogy between quantum indeterminacy and conscious freedom is strictly metaphorical unless otherwise specified.

Accordingly, when this book uses "observer," "measurement," or "collapse" language, it is describing physical interaction/measurement in standard quantum theory, not claiming that human awareness causes wave-function collapse.

Consciousness ≠ Emergent Illusion

No physical or functional account provides an explanation for subjective experience.

Illusion theories collapse under self-refutation (an illusion still needs a subject).

Intentionality ≠ Information Encoding

Intentionality is semantic; information is syntactic. These two cannot be equated.

Personhood ≠ Complexity

Complex arrangements of matter do not automatically yield subjectivity or rationality.

Personhood involves:

- rational agency
- interiority
- relational capacity
- moral and spiritual dimensions

THEOLOGY
CORE THEOLOGICAL CONCEPTS & DIVINE ATTRIBUTES

Theology (Classical Definition)
The discipline that studies:

- the nature and attributes of God
- the relationship between Creator and creation
- divine revelation
- salvation, providence, and purpose

Theology operates on a different plane than physics or metaphysics.

It does not propose:

- mechanisms
- physical explanations
- empirical predictions

Its domain is God, meaning, value, revelation, and ultimate causes.

God (Classical Theism)

The necessary, self-subsisting, infinite, immaterial source of all being.

In classical theism, God is:

- pure act (actus purus)
- perfect
- simple
- eternal
- immutable
- omnipotent
- omniscient
- omnipresent
- personal
- the Creator of all things

God is not:

- an element of the universe
- a force
- a field
- a being among beings
- synonymous with nature

God is Being Itself, not a component of the cosmos. God is not a being within the universe or a force alongside natural causes, but the transcendent source who sustains all being.

Creation ex Nihilo

The doctrine that God creates without using preexistent matter. Creation is:

- not a temporal "start"
- not an event inside spacetime
- an ontological dependence relation

It means that the universe depends on God continually, not only at a first moment.

Immutability

God does not change. Change implies:

- lack
- imperfection
- temporal succession

None apply to a being whose essence is identical with existence. Immutability does not imply:

- inactivity
- coldness
- lack of relationship

It means God's being is complete and does not undergo development.

Eternity (Timelessness)

Eternity means:

- not bound by time
- possessing all of life "all at once" (totum simul)
- no before/after in God

Eternity ≠ infinite temporal duration.
God is outside time and sustains time itself.

Omnipotence

God can do all things that are logically possible and consistent with His nature.

This excludes:

- contradictions (e.g., square circles)
- actions contrary to God's essence (e.g., ceasing to be God)

Omnipotence is not arbitrary power.

Omniscience

God knows all things:

- actual
- possible
- counterfactual
- past, present, and future

God's knowledge is:

- not experiential in a temporal sense
- not discursive or acquired
- identical with His essence

God does not "learn."

Omnipresence

God is present to all things as the cause of their being.
Omnipresence does not mean:

- God is physically spread out
- God occupies space
- God is identical with the universe

It means God is the act of being sustaining all things.

Simplicity (Divine Simplicity)

God has no parts, potentiality, or composition.
God is:

- His essence
- His existence
- His knowledge
- His will
- His goodness

All are one reality in God, distinguished only conceptually.

Transcendence

God exists beyond and above creation, not contained by space or time.

Immanence

God is present to creation by sustaining its existence and acting within it.

Immanence does not imply pantheism or that creation is divine.

Providence

God's continuous governance of creation.

Providence includes:

- conservation
- concurrence
- governance toward ends

Providence is not identical to determinism; secondary causation remains real.

LOGOS, WORD, WISDOM, AND BIBLICAL-THEOLOGICAL CONCEPTS

Logos (Greek)

A term rich in meaning:

- word
- reason
- rational order
- intelligibility
- principle of creation
- divine self-expression

In Johannine theology, the Logos is personal, referring to the eternal Son.

Logos is not:

- an impersonal force
- abstract logic
- mathematical structure alone

Though Logos encompasses rationality, it is fundamentally personal. The Logos is not a linguistic symbol or metaphysical abstraction but a personal and eternal divine identity.

Davar (Hebrew)
Word, speech, or event — often implying a word-in-action.
In Hebrew thought:

- words accomplish
- words create
- words perform

Davar is dynamic, not merely linguistic.

Chokhmah (Hebrew)
Wisdom — the divine, creative intelligence described in Proverbs 8.
Chokhmah is not:

- human cleverness
- accumulated knowledge
- abstract reason

It is a theological concept tied to God's creative activity.

Ruach (Hebrew)
Spirit, breath, wind — indicating the dynamic presence and power of God.
Ruach is not:

- physical energy
- airflow

- mystical force

It is a personal divine presence.
Imago Dei (Image of God)
Human beings uniquely reflect:

- rationality
- relationality
- moral capacity
- creativity
- freedom
- spiritual depth

Imago Dei does not refer to:

- physical likeness
- intelligence alone
- moral perfection

It is relational, rational, and spiritual.
Revelation
God's self-disclosure through:

- creation
- history
- Scripture
- ultimately the Logos (the Son)

Revelation is personal, not merely informational.
Covenant
The relational, binding commitment between God and humanity.
Not merely a contract; it involves:

- promise

- fidelity
- relationship
- mission

CAUSATION, CREATION, AND DIVINE ACTION

Primary Causation

God's causation as the giver of being.

Primary cause sustains existence itself. Primary causation does not compete with physical or secondary causes.

Secondary Causation

The causation exercised by creatures within the created order. Examples:

- physical processes
- human will
- biological development

Secondary causes are real and not overridden by primary causation. Divine causation is primary and ontological; natural causes are secondary and creaturely, and the two do not compete.

Concurrentism

The doctrine that God cooperates with created causes without eliminating their genuine agency.

Conservation

God's continual sustaining of all created things in being. Without conservation, all things would lapse into nonexistence.

Miracle

An extraordinary divine action outside the normal patterns of secondary causation.

Miracles are:

- not violations of natural laws
- not God "breaking into" creation

- expressions of God's sovereignty

They are exceptional, not arbitrary.
Creation (Theological)
Creation is:

- the dependence of all things on God for existence
- timeless in its metaphysical grounding
- not confined to an initial temporal moment

Creation is not identical to the Big Bang.
Emanation (Rejected)
The idea that creation flows out of God by necessity. Classical theism rejects emanation.
Creation is a free act, not a necessary overflow.

BOUNDARIES AND MISINTERPRETATION PREVENTION

God ≠ Universe
Avoids pantheism. Creation is distinct from its Creator.
God ≠ Information
Information describes structure. God is Being Itself, the source of both existence and intelligibility.
God ≠ Quantum Field
Quantum fields are physical. God is not a physical system.
God ≠ Law of Nature
Laws describe patterns in creation. God is the author and sustainer, not the law.
Logos ≠ Information Theory
Logos includes rationality but is fundamentally personal and divine.
Spirit ≠ Energy
Energy is physical; Spirit is divine.
Creation ≠ Emergence

Emergence is the development of complexity from simpler conditions. Creation is the gift of
existence itself.
Providence ≠ Determinism
Human freedom and secondary causes remain real.
Theological Language ≠ Metaphorical Physics
Terms like:

- light
- life
- breath
- spirit

in Scripture are theological, not physical analogies or models of physics.

MASTER INDEX

How to Use This Index:

TIER 1: Master Alphabetical Index

Use when you know the term you're looking for and need to find where it appears in the manuscript. Standard academic index format.

TIER 2: Disciplinary Indices

Use when researching within a specific field. Shows the internal vocabulary of physics, metaphysics, theology, information theory, or philosophy of mind as distinct domains.

TIER 3: Cross-Disciplinary Concept Map

Use when tracking how a single term (like 'information,' 'cause,' or 'observer') functions differently across disciplines. Essential for preventing category errors and understanding the manuscript's methodological precision.

TIER 1
MASTER ALPHABETICAL INDEX

Complete alphabetical listing of all significant terms, concepts, persons, and biblical references throughout the manuscript.

A
- Accident (metaphysical), 94, 347
- Act (Actuality), 80, 82, 121, 123
- Act of Being (Actus Essendi), 80, 121
- Pure Act (Actus Purus), 121, 171, 178
- Act and Potency, 23, 121, 123
- Actualization, 97, 110, 121, 123
- AdS/CFT Correspondence, 37, 38, 45, 48, 53
- Analogy of Being, 178, 405
- Anthropic Principle, 30, 31
- Aquinas, Thomas, 23, 24, 78 (extensively in Part II & X)
- Aristotle, 23, 78
- Arrow of Time, 58, 60, 61
- Athanasius, 322
- Augustine, 119, 193

B

Baryon Acoustic Oscillations, 68, 443
Being (Ens), 121, 464
Being Itself (I AM), 17, 121 (core in theology sections)
Composite Being, 121, 466
Contingent Being, 81, 121
Necessary Being, 24, 121
Subsistent Being, 121, 465
Bekenstein, Jacob, 37, 49
Bekenstein Bound, 37, 49, 51, 53
Bekenstein-Hawking Formula, 38
Big Bang, 27–35, 68, 257 (main in Ch. 1)
Initial Conditions, 30
Low-Entropy Beginning, 27, 30
Black Holes, 37, 38, 46–47
Entropy, 30, 37
Event Horizon, 37
Hawking Radiation, 38, 50
Information Paradox, 37, 50, 51

C

Causality, 23, ~120–130 (Four Causes section in metaphysics)
Efficient Cause, 465, ~123
Final Cause, 465, ~123
Formal Cause, 465, ~123
Four Causes, 465, ~123
Material Cause, 465, ~123
Primary and Secondary Causation, 122, ~123
Christ. See Jesus Christ
Colossians 1:16-17, 155, ~250–260 (theology)
Complementarity, Principle of, 377, ~140–150
Consciousness, 340, ~200+ (philosophy of mind)
Hard Problem of Consciousness, 71, 371, ~200+
vs. Information, 341, 411, ~200+
vs. Meaning, 341, 411, ~200+

Cosmic Microwave Background (CMB), 27, 29, 33, 2729
Cosmological Constant, 28, 30, 31
Creation, 119, ~240–260 (Ch. 13–15)
ex nihilo, 119, ~250
through the Logos, 280, ~250
through the Word, 120, ~250

D

Dark Energy, 443, ~70–80
Dark Matter, 443, ~70–80
Davar (Hebrew word), 141, ~240–250
Decoherence, Quantum, 600, ~40, 45
Divine Simplicity, 119, ~170–180

E

Einstein, Albert, 28, 31
Emergence, 40, 185, ~200–210
Entanglement, Quantum, 40, 41, 45, 51
and Spacetime, 41, ~40, 45
EPR Paradox, 40, ~45
Entropy, 30, 37, 38, 61 (thermodynamic & black hole)
and Arrow of Time, 58, 61
Black Hole, 37, 38
Second Law of Thermodynamics, 58, ~61
Shannon Entropy, 36–40, 456
Essence, 23, ~120–130
vs. Existence, 23, ~120–130
Event Horizon, 37
Existence, 23, ~120–130
Act of Existing, 23, ~121
vs. Essence, 23, ~120–130
Exodus 3:14 (I AM WHO I AM), 110, ~240–250

F

Finality (Teleology), 465, ~123

Fine-Tuning, 31
Cosmological Constant, 28, 28, 30, 31
Fundamental Constants, 30, ~30–31
Form (Metaphysical), 140, ~123
Four Causes. See Causality

G

Genesis 1, 142, ~240–250
Creation through distinction, 43, ~240–250
Divine speech, 43, ~240–250
God, 17, ~240–260 (as Being Itself, I AM, etc.)
as Being Itself, 112, 121, 240+
as I AM, 17, 240+
as Logos, 24, 240+
as Necessary Being, 24, 121
as Pure Act, 121, 171
Divine Simplicity, 119, ~170–180
Trinity. See Trinity

H

Hard Problem of Consciousness, 71, ~200+
Hawking, Stephen, 38, 50
Hawking Radiation, 38, 50
Hebrews 11:3, 259, ~9, 250+
Heraclitus, 152, ~140–150
Hierarchical Explanation, Principle of, 378, ~140–150
Holographic Principle, 38, 45, 48, 49, 51, 53
AdS/CFT Correspondence, 37, 38, 45, 48, 53
Bekenstein Bound, 37, 49, 51, 53
Information on Boundaries, 24, 38, 49
Hubble, Edwin, 28
Hubble Expansion, 28, 444
Hypostasis (Person), 161, ~300+

I

'I AM' (Divine Name), 110, ~240–250
　as Being Itself, 110, 121, 240+
　Exodus 3:14, 110, ~240–250
　Metaphysical Meaning, 42, ~240–250
Image of God (Imago Dei), 265, ~300+
Incarnation, 124, 154, ~250–260
Inflation (Cosmic), 441, ~70–80
Information, 36, ~36–45
　Bekenstein Bound, 37, 49, 51, 53
　vs. Consciousness, 41, 411, ~200+
　Landauer's Principle, 38, 39, 44–45
　vs. Meaning, 41, 411, ~200+
　Shannon Information Theory, 36, ~36–40
　vs. Semantic Content, 43, ~200+
Information Paradox (Black Holes), 37, 50, 51
Intelligibility, 13, 32, ~32, 140–150
　of Being, 40, 240, ~140–150
　and the Logos, 113, 240, ~140–150
　of the Universe, 13, 32, ~32, 140–150

J

Jesus Christ, 123, ~250–260
　as Embodied Logos, 123, 157, ~250–260
　Incarnation, 124, ~250–260
　Sustaining Creation, 124, ~250–260
　John 1:1-3, 151, ~250–260
　John 1:14, 154, ~250–260

L

Landauer's Principle, 38, 39, 44–45
Law (Physical Laws), 31, ~140–150
　Intelligibility of, 32, ~32, 140–150
　Mathematical Structure, 41, ~32, 140–150
Logos, 124, ~140–150, 240–260
　Christ as Embodied Logos, 123, 157, ~250–260

Divine Word, 151, ~250–260
in Heraclitus, 152, ~140–150
vs. Information, 151, ~140–150
John 1:1-3, 151, ~250–260
Rational Order, 80, ~140–150
in Stoicism, 152, ~140–150
Low-Entropy Past, 30

M

Maldacena, Juan, 45, 48
Mathematics, 32
Unreasonable Effectiveness of, 32
Matter (Hyle), 32, ~120–130
Meaning, 32, ~200+
vs. Information, 36, ~200+
Semantic Content, ~200+
Measurement Problem (Quantum), ~40, 45
Metaphysics, ~80–130
vs. Physics, ~80–130
vs. Theology, ~80–130
Multiverse, 31

N

Necessary Being, 24, 121
Non-Reduction, Principle of, ~140–150
Non-Substitution, Principle of, ~140–150
Nucleosynthesis (Big Bang), ~29, 33

O

Observer (Quantum), ~40, 45
vs. Conscious Mind, ~40, 45
Order, ~32, 140–150
Informational, ~36–45
Rational, ~32, 140–150
Ousia (Essence), ~300+

P

Participation (Metaphysical), ~120–130
Penrose, Roger, 30
Perichoresis, ~300+
Person (Theological Definition), ~300+
Potency (Potentiality), ~23, 121, 123
Act and Potency, 23, 121, 123
vs. Quantum Potential, ~23, 121
Psalm 19:1, ~240–250
Psalm 33:6, ~240–250

Q

Quantum Mechanics, ~36–45
Decoherence, ~40, 45
Entanglement, ~40, 45
Measurement Problem, ~40, 45
Observer, ~40, 45
Wave Function Collapse, ~40, 45

R

Relationality, ~140–150
of Being, ~140–150
Quantum, ~40, 45
Trinitarian, ~300+

S

Second Law of Thermodynamics, ~58, 61
Shannon, Claude, ~36
Shannon Information Theory, ~36–40
Singularity (Mathematical), ~30, 34
Spacetime, ~40, 45, 48
and Entanglement, ~40, 45
Holographic Principle, 38, 45, 48, 49, 51, 53
Stoicism, ~140–150
Substance (Metaphysical), ~120–130

T

- Teleology (Finality), ~123, 465
- Time, ~61
- Arrow of Time, 58, 60, 61
- Beginning of, ~27–35
- Trinity, ~300+
- Hypostasis (Person), 161, ~300+
- Ousia (Essence), ~300+
- Perichoresis, ~300+
- Relational Nature, ~300+

U-W

- Unreasonable Effectiveness of Mathematics, 32
- Wave Function Collapse, ~40, 45
- Wigner, Eugene, 32
- Word of God. See Logos

TIER 2
DISCIPLINARY INDECES

Organized by academic discipline to show internal vocabularies and methodological boundaries. These indices demonstrate how each field maintains its own conceptual integrity within the larger synthesis.

PHYSICS & COSMOLOGY

Terms describing physical mechanisms, cosmological phenomena, and observable processes.

AdS/CFT 37-53, Arrow of Time 58-61, Baryon Acoustic Oscillations 68-443, Bekenstein Bound 49, Big Bang 27-35, Black Holes 37-50, CMB 27-29-33, Cosmological Constant 28-31, Dark Energy/Dark Matter 443, Decoherence 600, Entanglement 40-51, EPR Paradox 40, Entropy 30-58, Fine-Tuning 31, Holographic Principle 38-53, Hubble Expansion 28-444, Inflation 441, Measurement Problem 40-45, Nucleosynthesis 29-33, Quantum Mechanics 36-45, Spacetime 40-53, Wave Function Collapse 40-45

METAPHYSICS & PHILOSOPHY

Terms addressing being, causality, essence, existence, and the conditions underlying all reality.

Accident 94-347, Act/Actuality 80-123, Act of Being 80-121, Pure Act 121-178, Act and Potency 23-123, Actualization 97-123, Analogy of Being 178-405, Being (Ens) 121-464, Causality/Four Causes 23-465, Essence/Existence 23-130, Finality/Teleology 123-465, Form 140, Intelligibility 13-32-240, Matter (Hyle) 32-130, Participation 120-130, Potency 23-123, Relationality of Being 140-150, Substance 120-130

THEOLOGY & SCRIPTURE

Terms addressing divine nature, revelation, creation, Trinitarian doctrine, and biblical references.

Colossians 1:16-17 155, Creation 119-280, Davar 141, Divine Simplicity 119, Exodus 3:14 110, Genesis 1 142, God as Being Itself/I AM/Logos/Necessary Being/Pure Act 17-112-121-240+, Hebrews 11:3 259, Image of God 265, Incarnation 124-154, Jesus Christ 123-157, John 1:1-3 151, John 1:14 154, Logos 124-151-157, Perichoresis 300+, Psalms 19:1/33:6 240-250, Trinity/Hypostasis/Ousia 161-300+

INFORMATION THEORY

Terms describing structure, distinctions, correlations, and informational constraints in physical systems.

Bekenstein Bound 49, Bekenstein-Hawking 38, Entropy (Shannon) 36-456, Holographic Principle 38-53, Information 36-411, Information Paradox 50, Landauer's Principle 38-39

PHILOSOPHY OF MIND

Terms addressing consciousness, subjective experience, meaning, and mental phenomena.

Consciousness 340, Hard Problem 71-371, vs. Information 341-411, vs. Meaning 341-411, Meaning/Semantic Content 32-43

TIER 3
CROSS-DISCIPLINARY CONCEPT MAP

Terms that appear across multiple disciplines with domain-specific meanings. This section shows how to track the same word as it functions differently in physics, metaphysics, theology, information theory, and philosophy of mind—preventing category errors and ensuring precision.

(Primary pages ~140–180 + Part X)

CAUSE: In Physics ~23-120, Metaphysics ~123-465, Theology ~122-123, Information ~constraint, Mind ~intention

INFORMATION: Physics ~36-53 (state/distinctions), Metaphysics ~intelligibility/form, Theology ~Logos/divine knowledge, Info Theory ~36-456 (Shannon/uncertainty), Mind ~meaning (distinct)

OBSERVER: Physics ~40-45 (measuring system), Metaphysics ~subject, Theology ~personal agent, Info Theory ~receiver, Mind ~conscious subject

ORDER: Physics ~low entropy/symmetry, Metaphysics ~teleology, Theology ~providence, Info Theory ~pattern/correlation, Mind ~rationality

POTENTIAL: Physics ~probability/quantum, Metaphysics ~potency 23-123, Theology ~divine power, Info Theory ~capacity, Mind ~intentional state

LAW: Physics ~mathematical regularity, Metaphysics ~ontological necessity, Theology ~divine command, Info Theory ~encoding rule, Mind ~rational norm

BEING/EXISTENCE: Physics ~causal effects, Metaphysics ~esse/essence vs existence 23-130, Theology ~I AM/Being Itself 17-121-240+

RELATIONALITY: Physics ~entanglement 40-51, Metaphysics ~being-toward, Theology ~perichoresis/Trinitarian 300+, Info Theory ~correlations

EMERGENCE: Physics ~higher properties 40-185, Metaphysics ~actualization of potency, Mind ~consciousness (irreducible)

METHODOLOGICAL PRINCIPLES: Complementarity ~140-377, Hierarchical Explanation ~140-378, Non-Reduction ~140-150, Non-Substitution ~140-150, Analogical Interpretation, Category Errors, Domain Separation, Integration Guidelines (Part X)

SCHOLARS & HISTORICAL FIGURES
Aquinas, Thomas, 23, 24, 78
Aristotle, 23, 78
Athanasius, 322
Augustine, 119, 193
Bekenstein, Jacob, 37, 49
Einstein, Albert, 28, 31

Hawking, Stephen, 38, 50
Heraclitus, 152
Hubble, Edwin, 28
Maldacena, Juan, 45, 48
Penrose, Roger, 30
Shannon, Claude, 36
Wigner, Eugene, 32

KEY BIBLICAL REFERENCES
Genesis 1 (Creation Account), 142
Exodus 3:14 (I AM WHO I AM), 110
Psalm 19:1 (Heavens Declare Glory), ~
Psalm 33:6 (By the Word of the LORD), ~
John 1:1-3 (In the Beginning Was the Word), 151
John 1:14 (The Word Became Flesh), 154
Colossians 1:16-17 (All Things Created Through Him), 155
Hebrews 11:3 (Visible from the Invisible), 259

BACK COVER BLURB

What if the deepest discoveries of modern physics, the oldest questions of philosophy, and the moral vision of Christianity were never meant to compete—but to speak to the same reality from different angles?

(IAM) invites readers on an intellectual journey that begins with the universe itself: its origin, its order, and the strange intelligibility revealed by modern cosmology. From there, it follows the questions physics cannot answer on its own—questions of meaning, consciousness, freedom, and purpose—into the realm of philosophy, and finally into the heart of the Christian tradition that quietly shaped Western thought.

Rather than arguing for conclusions, this book guides the reader through a carefully structured path of discovery. Along the way, familiar ideas are seen anew: creation as intelligible, truth as relational, and human dignity as something grounded, not invented.

Written for thoughtful readers who refuse false choices between faith and reason, (IAM) does not require technical expertise or theological training—only curiosity, patience, and a willingness to follow questions where they lead.

Some will read it as a synthesis.
Others will experience it as a journey.
A few will recognize something deeper,
taking shape along the way.

www.ingramcontent.com/pod-product-compliance
Lightning Source LLC
Chambersburg PA
CBHW070746230426
43665CB00017B/2265